Factors in Formation and Regression of the Atherosclerotic Plaque

NATO ADVANCED STUDY INSTITUTES SERIES

A series of edited volumes comprising multifaceted studies of contemporary scientific issues by some of the best scientific minds in the world, assembled in cooperation with NATO Scientific Affairs Division.

Series A: Life Sciences

Recent Volumes in this Series

This series is published by an international board of publishers in conjunction with NATO Scientific Affairs Division

A	Life Sciences	Plenum Publishing Corporation
B	Physics	London and New York
C	Mathematical and Physical Sciences	D. Reidel Publishing Company Dordrecht, The Netherlands and Hingham, Massachusetts, USA
D	Behavioral and Social Sciences	Martinus Nijhoff Publishers The Hague, The Netherlands
E	Applied Sciences	

Factors in Formation and Regression of the Atherosclerotic Plaque

Edited by
Gustav R. V. Born
University of London King's College
London, England

and

Alberico L. Catapano
Rodolfo Paoletti
Institute of Pharmacology and Pharmacognosy
University of Milan
Milan, Italy

PLENUM PRESS • NEW YORK AND LONDON
Published in cooperation with NATO Scientific Affairs Division

Library of Congress Cataloging in Publication Data

Main entry under title:

Factors in formation and regression of the atherosclerotic plaque.

(NATO advanced study institutes series. Series A, Life sciences; v. 51)
"Proceedings of a NATO Advanced Study Institute on the Formation and Regression of the Atherosclerotic Plaque, held September 3-13, 1980, in Belgirate, Italy" — Verso t.p.
"Published in cooperation with NATO Scientific Affairs Division."
Bibliography: p.
Includes index.
1. Atherosclerosis — Congresses. 2. Atherosclerosis — Etiology — Congresses. I. Born, G. R. V. II. Catapano, Alberico L. III. Paoletti, Rodolfo. IV. NATO Advanced Study Institute on the Formation and Regression of the Atherosclerotic Plaque (1980): Belgirate, Italy) V. North Atlantic Treaty Organization. Scientific Affairs Division. VI. Series. [DNLM: 1. Arteriosclerosis — Etiology — Congresses. 2. Arteriosclerosis — Metabolism — Congresses. WG 550 N2798f 1980]
RC692.F3 1982   616.1'36   82-10132
ISBN-13: 978-1-4684-4270-0   e-ISBN-13: 978-1-4684-4268-7
DOI: 10.1007/978-1-4684-4268-7

Proceedings of a NATO Advanced Study Institute on the Formation and Regression of the Atherosclerotic Plaque, held September 3-13, 1980, in Belgirate, Italy

PREFACE

Interest in the field of atherosclerosis research has broadened in recent years. However the main focus remains on the physiopathology of the arterial wall and on its interaction with blood constituents.

The purpose of this NATO Advanced Study Institute on "Factors in Formation and Regression of the Atherosclerotic Plaque" was to discuss the following points:

a) The physiopathology of the arterial wall;

b) Animal models;

c) Methods of studying the progression and regression of atherosclerotic lesions quantitatively;

d) The role of lipoproteins, platelets, smoke, alcohol, etc. in the formation of atherosclerotic lesions;

e) The pharmacological and dietary control of "risk factors."

This volume is a collection of the most relevant presentations on these topics. We hope it will provide a background for young scientists as well as a stimulus for further research to biologists and clinicians.

We wish to acknowledge the support of NATO and the Nutrition Foundation of Italy in organizing this Advanced Study Institute.

Gustav R. V. Born
Alberico L. Catapano
Rodolfo Paoletti

CONTENTS

PHYSIOLOGY AND PATHOLOGY OF THE ARTERIAL WALL

ANIMAL MODELS

LIPOPROTEINS AND ATHEROSCLEROSIS

PLATELETS AND ATHEROSCLEROSIS

OTHER FACTORS IN PROGRESSION AND REGRESSION

MORPHOLOGY AND PHYSIOLOGY OF THE ARTERIAL WALL

Giorgio Weber

Institute of Pathologic Anatomy, Center of Research on

Atherosclerosis, University of Siena, Siena, Italy

I) STRUCTURAL DIFFERENCES OF ARTERIES IN DIFFERENT BODY DISTRICTS

Correlations between Anatomy and Physiology on one side, Pathology of the wall and of the organs on the other are not well developed. It's common clinical observation in our countries that myocardial infarction concerns younger people and cerebral infarction
older people. We also know that different risk factors may be at the
basis of these age difference or different pathological lesions at
arterial wall level or even different structural and functional characteristics of these arterial districts.

In order to give a morphological basis to these and other similar
clinical observations, a pluridistrictual morphological study is required, extended to all or most of the arterial tree,a task which is
only at its beginning in humans as well as in most common animal models. In effect though in the last century and also more recently
comparisons of lesions and their degree have been performed, for
instance on superior and inferior limb arteries, not much is known
in structural differences which could induce also different pathological patterns of lesions. Only from 1977 myocardial infarct and cerebral infarct experts began meeting regularly and exchanging their
experiences! I do think that comparison of orthologic structures and
pathologic lesions in different body districts will give results in
the future. In Italy, a National Research Council Special Program is
on since four years.

Morbid Anatomic (postmortem) observation allows comparison among
structure and lesions in different body districts: two figures, from

1

the autopsy of a 22 years old woman dead because of a sub-arachnoidal
hemorrhage (Siena,aut 8759/12), show the descending branch of the
left coronary artery and the left middle cerebral artery: in the
coronary the" diffuse intimal thickening" (DIT) is well evident,
in the cerebral artery no signs of diffuse intimal thickening are
found. These different features in coronary and cerebral arteries
can well explain the earlier development of arterial lesion in the
coronaries and the earlier appearance of organ lesions in the heart.

The question is anyway still unsettled, as you know, on the
significance of DIT which has been studied so far practically only
at heart coronary level(review by Geer and Haust,1972 and more recently by Haust,1980) and to which a significant contribution has
been given by D. and C. Velican (1978). Little information is available in recent literature on the histology of diffuse intimal thickening at different ages in normal aorta not only in man but also in
non-human primates or in other animal models, actually used in atherosclerosis research.

Some observations are collected in a study by Stout and Thorpe
(1980) on the aorta of non-human primates where the DIT did not appear prevalent or more evident in any particular aortic area, so
that its distribution did not provide any clue as to its etiology
and as to its eventual pathogenic implications. At any rate, the
growth of DIT resulted accomplished by the addition of smooth muscle cells (SMCs) at the intimal medial junction.

It is not clear at all if and how those findings may be related
to the "intimal SMC masses" observed in the abdominal aorta of young
swine by Scott et al.(1979). Those masses are located predominantly
away from the blood vessel orifices; they appear more frequently
dorsally located in distal half of the abdominal aorta, ventrally
in the proximal half. The Authors observed early appearance of atherosclerosis (ATS) experimental lesions in corrispondence of those
cushions of SMCs. More subtle differences between aortic and abdominal
segments have been pointed out by Goldberg et al.(1980) who observed
greater myointimal proliferation in abdominal than in thoracic aorta
in rabbits subjected to intimal injury: it must be noted that the
regional variations observed in response to injury take place in
SMCs "which are phenotypically similar". And let us remind here that
Haley et al.(1977) documented at least two different populations of
SMCs in atherosclerotic lesions.

In cerebral arteries, areas of intimal muscular proliferation,
focally located near the orifices of the large distributing arteries,
have been described by Stehbens (1960) by means of serial arterial

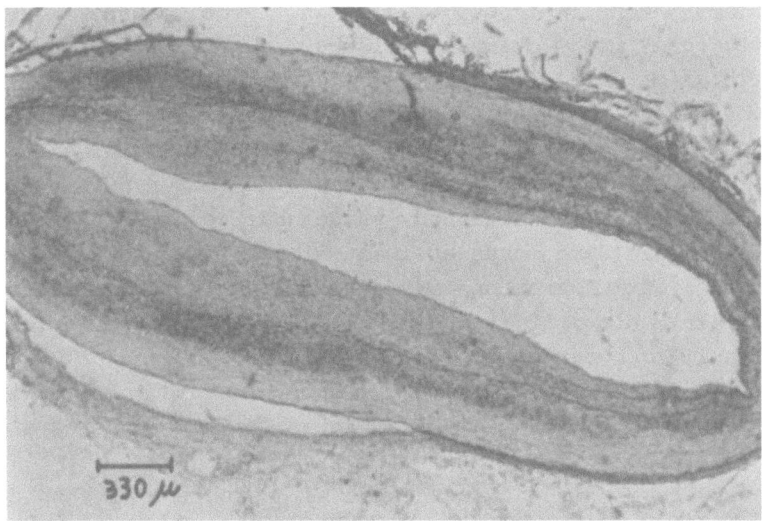

1

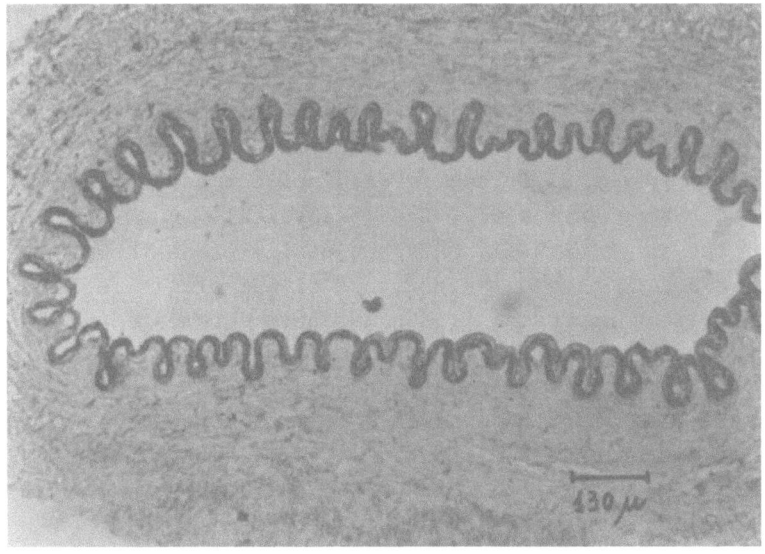

2

Fig. 1-2. Woman 22 years old (aut. 8759/12 Siena): no sign of diffuse
          intimal thickening (well evident instead in the left descend-
          ing coronary artery,2) in the left cerebral artery (1).

sections of human fetuses and infants. Those intimal pads,which appear regularly, were preceeded by elastic tissue changes and were constituted by musculo-elastic tissue.

On he significance of "DIT" (and other similar findings) there has been much speculation and some Authors consider those as early stages of atherosclerosis or as integral parts of athero-sclerosis; some Authors instead assume that they may represent "normal" (may be hemodynamically conditioned) structural components of an artery I wonder whether, in some cases at least, these findings could not represent the effect of previous adaptations for instance to preceeding periods of hypertension: it is known, in fact, that the arterial wall may adapt itself to hypertension through hypertrophy and/or hyper-plasia chiefly of SMCs (which are prevalently located in the intima at aortic level, in the media at renal artery level)(Limas et al., 1980).

While many people are now concerned with problems of morpho-metric quantitation of lesions and of optimizing selection of tissue blocks for histologic studies (cfr. McMahan et al.,1978), I feel that such findings and others (also at ultrastructural level) in normal looking arteries should not be underestimated, chiefly when initial lesions are studied because focal alterations of the normal structure are surely of relevance at the beginning of the ATS process.

The importance of the SMCs in the arterial wall is extremely high as generally admitted: changes in the activity of vascular enzymes (which are different at different levels) reflect corresponding changes of SMCs (Zemplenyi,1968), the metabolically most active component of the arterial wall. SMCs are easily prone therefore also to necrosis also beyond the presence of pathologic lesions: let us remember that necrosis of SMCs has been documented by Joris and Majno (1974) also in normal conditions. The importance of necro-tic changes should not be underestimated because they could be acting as a stimulus (at minor or major level) to proliferation of nearby viable SMCs of the arterial wall and/or evoking emargination (and attraction into the intima) of mononuclear cells whose presence has also been described in "normal" conditions by Joris et al. (1979) in the arterial wall.

SMCs being most susceptible to injurious agents, highly vulner-able (Constantinides,1965; Gillman,1967), "those arterial segments that contain abundant SMCs are eminently susceptible to atherosclero-sis" (Zemplenyi,1968) and could also be less susceptible to athero-regression.

Also morphogenetic and metabolic activities of the arterial SMCs, such as those recently studied by Hadjiisky et al.,1978) may show

segmental differences: in younger rats for instance greater morphogenetic activity has been found in the aortic SMCs, higher contractile differentiation in the femoral SMCs.

Differences are not only bound to topography; age and sex differences have been recently statistically evaluated at autopsy by the
Velicans (1980)(in press), thus confirming previous informations
by different Authors (cfr. Dock,1946).

It is someway astonishing that sex differences in proliferation
of vascular SMCs have been observed even in primary cultures of cells
obtained from 10 weeks-old rats: cells taken from male animals do
proliferate more quickly than those from females (Travo et al.,1980).

Differences may also depend on seasonal variations in the bio-
physical properties of rabbit aorta and therefore of its susceptibility to experimental atherosclerosis, as observed by Oxlund et al.
(1979) who had previously noted (Helin et al.,1969) a relative im-
munity to atherosclerosis of rabbits during the hair-shedding period.

Differences depend also on the animal species: not to say of
the well known differences among thoracic and abdominal aorta in
birds, it may be remembered here that,at histochemical level Hadjiisky et al. (1974) observed a lack of 5-nucleotidase activity
(already demonstrated by Antonini and Weber,1951, in rabbits and
man) in the aorta of dogs and rats, which are species "resistant"
to experimental atheroslerosis. This activity is instead present
in pigs and in rabbits, species wich are both prone to develop
experimental atherosclerosis. It has been recently shown also that
the aortic chondroitin 6-sulphate content is higher in animal species
which are susceptible to ATS (Toledo and Mourao,1980).

More examples could be quoted here; a large quantity of informations can also be collected for instance from the Heidelberg 1973
Conference, ed. by Wolf and Werthessen(1975).

But let us look now at the ECs. Shape differences of endothelial cells (ECs) (even of great relevance) may appear at different
levels of the vascular tree: they have been studied by Buss et al.
(1979) at scanning electron microscopy (SEM)in rabbit veins and by
Kibria et al. (1980) in pulmonary trunk, pulmonary veins, aorta and
inferior vena cava.

Inter-endothelial junctions may show different ultrastructural
patterns at different levels of the arterial tree (Huttner and Peters,1978; Simionescu et al.,1976; Simionescu,1977,1980) when studied
by means of transmission electron microscopy (TEM) and freeze-etching
(FE) techniques. It is since long time well known that great topo-
graphic differences in pigs aorta can be observed: areas of spontaneously occurring enhanced permeability to ^{131}I-albumin and

[131]I-fibrinogen (Bell et al.;1974,a,b) have been demarcated by their
uptake of the protein binding azodye Evan blue (Bell et al.,1972).
These areas exhibit differences also in ECs morphology (Gerrity
et al.,1977).

Increased ECs turnover (Caplan and Schwartz,1973) has also been
observed by those Authors,confirming previous observations by Payling
Wright (1971) in Guinea pigs and by Magnani and Coccheri in rabbits
(1960). The areas of dye uptake are more susceptible (Gerrity et al.,
1976) to endotoxin induced endothelial injury. Gerrity et al.(1979)
demonstrated also monocyte adherence to the endothelium in "blu
areas" in 86% of samples (52% in white areas) after short-term
cholesterol feeding (only 17% in controls). Also the number of
monocytes within the intima was higher in "blu areas" during early
atherogenesis and still higher thereafter. The foam cells in the lesion
appeared to be completely derived from blood-borne monocytes (reviewed by Stary at the II International Austrian Conference, Vienna Meeting, 1980).

Those different structural aspects of the endothelial layer
and other aspects that are going to be studied suggest that still
more variants of morphological basis of function may be found.
Problems of permeability have been recently reviewed by Walton
(1980).

II) THE ENDOTHELIUM AS A "FUNCTIONALLY ACTIVE COMPONENT OF THE
    ARTERIAL WALL"

Since the classic monograph by Altschul (1954), the review by
Haust (1977) and the lecture held by Gimbrone Jr. at the recent
Houston V International Symposium (1980), we have been more and more
realizing that the integrity of this "structurally simple but functionally complex" layer is essential to the health of the arterial
wall.

The biochemistry of EC surface of capillaries has also progressed up to revealing the existence of a preferential distribution of
anionic sites and monosaccharyde residues (Simionescu,1980): FC surface properties such as the distribution and movement of anionic
membrane components may be relevant for the normal function of the
blood vessel wall and its involvement in disease process such as
thrombosis, inflammation and ATS (Pelikan et al.,1979).

After Gospodarowicz (1980), actively growing vascular ECs in-
ternalize and degrade LDL via a receptor-mediate pathway which
regulates the cell metabolism of cholesterol and the number of
surface receptors site for low density lipoproteins (LDL). Highly

contact-inhibited endothelial monolayer brings about a complete
inhibition of LDL internalization and lysosomal degradation. When
contact inhibition in endothelial monolayer is locally released
by wounding (Vlodavsky et al.,1978) cells released from contact-
inhibition accumulate LDL-cholesterol. Reversal of contact-inhibi-
tion by wounding provides a mechanism by which we understand that
the endothelium could really represents the primary initiation of
the ATS lesions (Vlodavsky et al.,1978).

We know today that lipoprotein-lipase, which hydrolyses the
triglyceride component of plasma chylomicrons, appears to be bound
to the glycoprotein Con-A reactive coat evidenced by Weber et al.
(1972;1973) at the luminal endothelial surface (Olivecrona et al.,
1976; Dicorleto and Zilversmit,1975), even if we still don't exactely
know (Wissler,1980) the precise function of the "glycocalyx" in the
atherogenic process: it was observed with different techniques that
the glycocalyx may show thickness variations during early phases
of experimental atherogenesis (Weber,1973; Jellinek,1974) and variations from normal to focal absence have been more recently observed
also in advanced aortic intimal lesions of hypertensive rats (Limas
et al.,1980).

Endothelium possesses not only membrane associated lectin
binding sites (cfr. Stein et al.,1976) but also specific receptors
for LDL (Stein and Stein,1976; Vlodavsky et al.,1978),for insulin
(Gimbrone and Alexander,1977), vasoactive hormones (Buonassisi and
Venter,1976), histocompatibility (Gibofsky et al.,1975; Moares and
Stastny,1977), blood group antigens (Jaffe et al.,1973), proteolytic
activity and inhibitions (Tokes and Sorgente,1976; Becker and Harperl,
1976), enzymes important in lipid (Blanchette-Mackie and Scow,1971)
and vasoactive peptide metabolism (Ryan et al.,1976; Pelikan et al.,
1979).

In vitro, it has been shown (Gospodarowicz,1980) that contact-
inhibition of ECs is associated with sharp decreases in Con-A reactivity and LDL surface receptor lateral mobility and correlated with
accumulation of fibronectin (fibronectin is involved in a mechanism
by which the endothelium functions as a selective barrier to the
circulating levels of LDL in plasma)(C.J. Fielding et al.,1979;
P.E. Fielding et al.,1979).

An endothelial cells derived growth factor (ECDGF) has been
recently described by Gajdusek et al.(1980) with several features
differing from those of the platelet derived growth factor (PDGF).

So many endothelial functions are coming to light, that I think
we have completely to agree with Gimbrone(1980): our working concept

of endothelial injury which is fundamental in the Ross and Glomset's
(1976) "response-to-injury hypothesis" of atherogenesis "needs to be
broadened to consider endothelial integrity".

Endothelial integrity may be affected by chemical mutagens action
or viral transformation of endothelial cells whose surface is then
highly reactive to platelets in vitro and could bring also in vivo
to adhesion of platelets and to release of platelet products into
vessel wall.

The very thin cellular membrane which constitutes the endothelial layer does not clearly show, even at transmission electron microscopic examination, the basis of its functional complexity which
may be grouped chiefly in three aspects : <u>Selective permeability</u>
<u>barrier</u>, <u>Synthetic-metabolic-secretory-activities</u> and <u>Blood compati-</u>
<u>ble container</u> (reviewed by Gimbrone,1980).

On the basis of the data now available (and which will certainly
increase in the future), we can assume that not only "injuries" (anatomical discontinuities and more subtle, for instance, ultrastructural
lesions) but also "lack of integrity" may explain at the endothelial
level, abnormalities which can be somehow connected with the pathogenesis of atherosclerotic lesions.

As a Pathologist I would like to present you some observations
of ours concerning examples of endothelial "injury" and others which,
in my mind, could be assumed to represent morphological aspects of
endothelial "dysfunction" (cfr. Weber et al., 1978a; Weber et al.,
1980,in press).

Scanning and transmission electron microscopy studies have shown
us that, during experimental cholesterol atherogenesis (and/or after
immunological type injuries) rabbit aortic areas may be very precociously damaged. Necrotic lesions of single ECs or of groups of ECs,
involving the whole cellular body or only parts of it, may be observed (Weber et al.,1980) together with edematous reaction and
SMCs necrosis in underlying media. Microthrombi constituted by platelets (still granulated on the top, depleted in the depth) may be
found over those areas. In nearby areas you may find platelets
which are "creeping" beneath groups of partially detached, still
viable or necrobiotic ECs. Still viable ECs may be found, single or
in groups, in the peripheral arterial blood (Weber et al.,1978b;1979).
In other areas, single SMCs are lying at the interface among blood
stream and arterial wall without any sign of platelet adhesion of them
as for instance observed also at the surface of transplanted endothelial denuded aortas in pigs by Fuster et al.(1979).

If during experimental atherogenesis examples of ECs "injury"

can easily be found, I think that models of "lack of integrity"
ECs are to be found during atheroregression, if we look at
morphological ultrastructural features of ECs regenerating over
residual lesions: they look vermicular, plump, small, loosely adhering among themselves and to the underlying surface and, when more
tightly joining, their interendothelial junctions appear still loose
at SEM and straight at TEM; very few pinocytotic vesicles appear
at the luminal side of these small ECs. The Con-A reaction results
frequently positive not only at the luminal but even at the abluminal
side of their cellular membrane, which prospects to the enlarged
sub-endothelial space: where basement membrane debris and edematous
fluid are also frequently found (Weber et al.,1978b; 1980,in press).
Altered endocytotic process and possibly altered proteoglycan production may be signs of altered function in regenerating ECs (Davies
and Ross,1978; Davies et al.,1979; Vlodavsky et al.,1978).

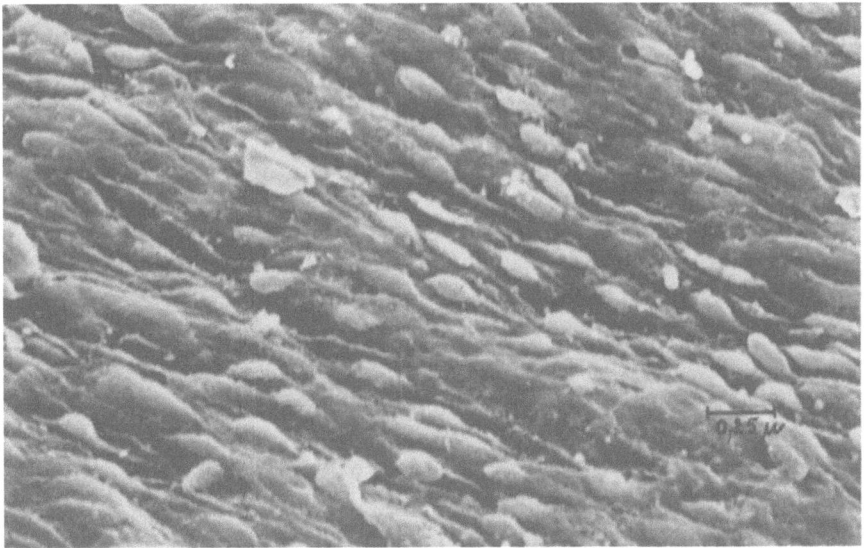

Fig. 3. Rabbit aorta during atheroregression: on the intimal sur-
        face, small plump regenerating, not yet well adjoining,
        endothelial cells. Scanning electron microscope Super ISI
        Mini Sem.

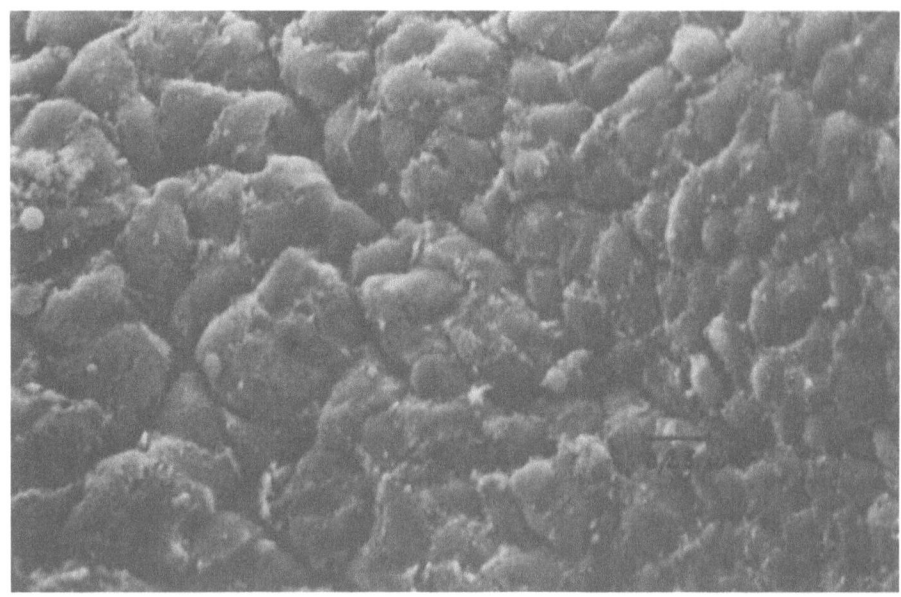

Fig. 4. Rabbit aorta during atheroregression: the intimal surface
        is covered by endothelial cells, only loosely adjoining
        to each other. Scanning electron microscope Super ISI Mini
        Sem.

     Recalling also the observations and suggestions of Christensen
et al. (1979a,b) and of Buck (1979a,b) on the subendothelial layer
structural characteristics and relationships with the endothelial
layer, we may infer that regenerating ECs in our "quick" models
are still in a dysfunctional condition. ECs dysfunction could explain
also the observations by Minick et al.(1977) and by Falcone et al.
(1980) after extensive arterial denudation through balloon catheter-
ization (intimal thickening and lipid accumulation more marked in
areas covered by regenerating endothelium). But also apart from those
situations, also intimal areas exist in which due to the physiological
blood flow injury or other experimental subliminal injuries (Jorgensen et al.,1972; Nelson et al.,1976; Richardson and Moore,1980) the
"normal" conditions are always on the brink of pathology: those areas
are the circumstantial ones which, as well known, are so precociously
and frequently involved by ATS and where not seldom ECs are present
which show characteristics similar to the ones described above. In
our mind, also those cells should be interpreted as not perfectly

functioning; as also the circumostial intraparietal platelet accumulation recently described by Amstrong et al.(1980) seems clearly to
confirm (cfr. also recent review by Mason and Balis (1980).

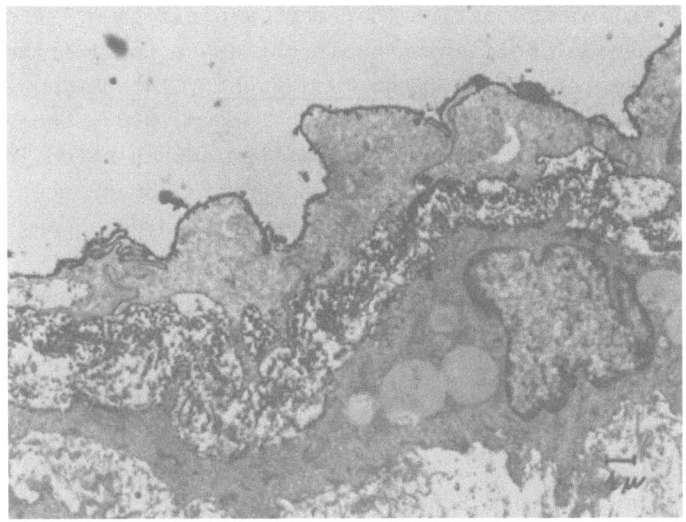

5

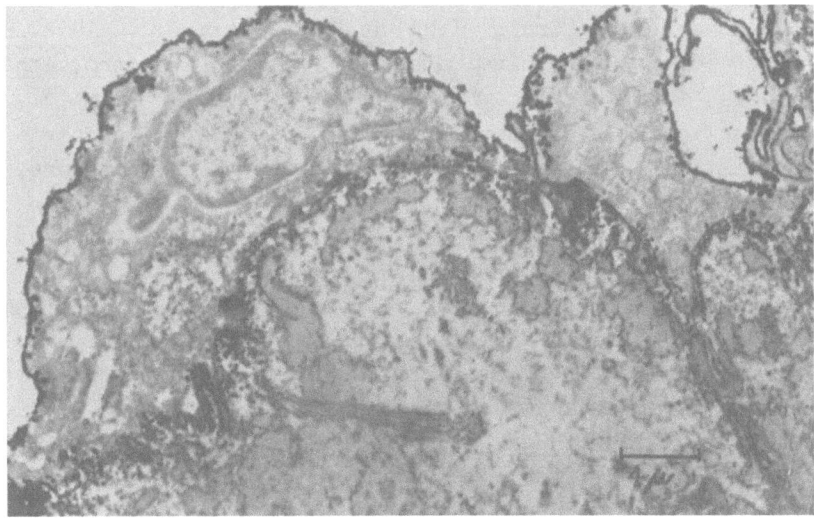

6

Figs. 5-6. Rabbit aorta during atheroregression: the small endothelial cells at the intimal surface show a very evident
Con-A positive reaction also at the abluminal cellular
membrane. Transmission electron microscope Siemens
Elmiskop 1 A.

REFERENCES

Altschul, R., 1954,"Endothelium, its development, morphology, func-
    tion and pathology",Mac-Millan Co., New York

Antonini, F.M. and Weber, G., 1951, Fosfatasi specifiche (5-nucleo-
    tidasi, ATP-pirofosfatasi) e fosfatasi aspecifica nella parete
    arteriosa normale, nell'arteriosclerosi umana, nell'arteriopatia
    sperimentale adrenalinica, Arch."De Vecchi", 14:985

Amstrong, M.L., Peterson, R.E., Hoak, J.C., Megan, M.B., Cheng, F.H.,
    and Clarke, W.R., 1980, Arterial platelet accumulation in experi-
    mental hypercholesterolemia, Atherosclerosis, 36:89

Aschoff, L., 1924, Atherosclerosis, in "Lectures in Pathology",
    Hoeber, New York

Becker, C.G. and Harpel, P.C., 1976, α_2-macroglobulin on human va-
    scular endothelium, J.Exp.Med., 144:1

Bell, F.P.,Adamson,I.L.,and Schwartz, C.J., 1974a, Aortic endothelial
    permeability to albumina. Focal and regional patterns to uptake
    and transmural distribution of ^{131}I-albumin in the young pig,
    Exp.Mol.Path., 20:57

Bell, F.P., Gallus, A.S.,and Schwartz, C.J., 1974b, Focal and regional
    patterns of uptake and the transmural distribution of ^{131}I-fi-
    brinogen in the pig aorta in vivo, Exp.Mol.Path., 20:281

Bell, F.P., Somer, J.B., Craig, I.H., and Schwartz, C.J., 1972, Pat-
    terns of aortic Evans blue uptake in vivo and in vitro, Athero-
    sclerosis, 16:369

Blanchette-Mackie, E.J.,and Scow, R., 1971, Sites of lipoprotein li-
    pase activity in adipose tissue perfused with chylomicrons,
    J.Cell Biol., 51:1

Buck, R.C., 1979a, Contact guidance in the subendothelial space. Re-
    pair of rat aorta in vitro, Exp.Mol.Path., 31:275

Buck, R.C., 1979b, The longitudinal orientation of structures in the
    subendothelial space of rat aorta, Am.J.Anat., 156:1

Buonassisi, V. and Venter, J.C., 1976, Hormones and neurotransmitter
    receptors in an established vascular endothelial cell line ,
    Proc.Natl.Acad.Sci.(USA), 73:1612

Buss, H., Schneider, J., and Hollweg, H.G., 1979, The endothelial
    surface of large veins of rabbits: scanning electron microscopic
    observations, Path.Res.Pract., 165:392

Caplan, B.A. and Schwartz, C.J., 1973, Increased endothelial cell
    turnover in areas of in vivo Evans blue uptake in the pig aorta,
    Atherosclerosis, 17:401

Christensen, B.C., Chemnitz, J., Tkocz, J., and Kim, C.M., 1979a, Repair in rarterial tissue. I. Endothelial regrcwth.Subendothelial tissue changes and permeability in the healing rabbit thoracic aorta, Acta Path.Microbiol.Scand. Section A, 87:265

Christensen, B.C., Chemnitz, J., Tkocz, J., and Kim, C.M., 1979b, Repair in arterial tissue.II. Connective tissue changes following an embolectomy catheter lesion. The importance of the endothelial cells to repair and regeneration, Acta Path.Microbiol. Scand. Section A, 87:275

Constantinides, P., 1965, "Experimental Atherosclerosis", Elsevier, Amsterdam

Davies, P.F. and Ross, R., 1978, Mediation of pinocytosis in cultured arterial smooth muscle and endothelial cells by platelet-derived growth factor, J.Cell Biol., 79:663

Davies, P.F., Seldom, S.C., and Schwartz, S.M., 1979, Enhanced rates of fluid pinocytosis during exponential growth and monolayer regeneration by cultured arterial endothelial cells, J. Cell Phys., in press

Dicorleto, P. and Zilversmit, D.B., 1975, Lipoprotein lipase activity in bovine aorta, Proc.Soc.Exp.Biol.Med., 148:1101

Dock, W., 1946, The predilection of atherosclerosis for the coronary arteries, J.Am.Med.Assn., 131:865

Falcone, D.J., Hajjar, D.P., and Minick, C.R., 1980, Enhancement of cholesterol and cholesteryl ester accumulation in re-endothelialized aorta, Am.J.Path., 99:81

Fielding, C.J., Vlodavsky, I., Fielding, P.E., and Gospodarowicz, D., 1979, Characteristics of chylomicron binding and lipid uptake by endothelial cells in culture, J.Biol.Chem., 254:8861

Fielding, P.E., Vlodavsky, I., Gospodarowicz, D., and Fielding, C.J., 1979, Effect of contact inhibition on the regulation of cholesterol metabolism in cultured vascular endothelial cells, J.Biol. Chem., 254:749

Fuster, V., Bowie, E.J.W., Josa, M., Kaye, M.P., and Fass, D.N., 1979, Atherosclerosis in normal and Von Willebrand pigs: cross-aortic transplantation studies, VII International Congress Thromb. Haem., London, July 20, Comm. 1018- p. 425

Gaidusek, C., Dicorleto, P. Ross, R., and Schwartz, S.M., 1980, An endothelial cell-derived growth factor, J.Cell Biol., 85:467

Geer, J.C. and Haust, M.D., 1972, Smooth muscle cells in atherosclerosis, in "Monographs on Atherosclerosis", O.J. Pollack, H.S. Simms and J.E. Kirk,eds.,Springer Karger, Basel-Munchen-Paris-London-New York-Sidney

Gerrity, R.G., Naito, H.K., Richardson, M., and Schwartz, C.J., 1979,
    Dietary induced atherogenesis in swine : morphology of the intima
    in prelesion stages, Am.J.Pathol., 95:775
Gerrity, R.G., Richardson, M., Caplan, B.A., Cade, J.F., Hirsh, J.,
    and Schwartz, C.J., 1976, Endotoxin-induced vascular endothelial
    injury and repair. II. Focal injury, en face morphology, ^{3}H thy-
    midine uptake and circulating endothelial cells in the dog,
    Exp.Mol.Path., 24:59
Gerrity, R.G., Richardson, M., Somer, G.B., Bell, F.P., and Schwartz,
    C.J., 1977, Endothelial cell morphology in areas of in vivo Evans
    blue uptake in the aorta of young pigs. Part 2.Ultrastructure
    of the intima in areas of differing permeability to protein,
    Am.J.Path., 89:313
Gibofsky, A., Jaffe, E.A., Fotino, M., and Becker, C.G., 1975, The
    identification of HL-A antigens on fresh and cultured human
    endothelial cells, J.Immunol., 115:730
Gillman, T., 1967, Possible significance of arterial hyperplasia
    òr growth followed by involution in the genesis of arterial
    degeneratory disease, in "Le rôle de la paroi arterielle dans
    l'athérogénese" , L. Scebat,ed., Paris
Gimbrone, M.A., 1980, Endothelial dysfunction and the pathogenesis
    of atherosclerosis, in "Atherosclerosis V",A.M. Gotto, L.C.
    Smith and B. Alleh,eds., Springer Verlag, New York-Heidelberg-
    Berlin
Gimbrone, M.A. and Alexander, R.W., 1977, Insulin receptors in cul-
    tured human vascular endothelial cells, Circulation, 55:209
Goldberg, I.D., Stemerman, M.B., Ransil, B.J., and Fuhro, R.L., 1980,
    In vivo aortic muscle cell growth kinetics.Differences between
    thoracic and abdominal segments after intimal injury in the
    rabbit, Circulation Res., 47:182
Gospodarowicz, D., 1980, Interaction of LDL with the vascular endo-
    thelium, Second International Austrian Atherosclerosis Conf.,
    Vienna 7-12 April(Abstract presented but not read)
Hadjiisky, P., Jurukova, Z., Renais, J. ,and Scebat, L., 1978, Seg-
    mental differences in morphogenetic activity of arterial smooth
    muscle cells. Histochemical and radioautographic studies ,
    Connective Tissue Res., 6:73
Hadjiisky, P., Renais, J., and Scebat, L., 1974, Etude comparativ
    des enzymes aortiques de rat (athérorésistant) et de lapin
    (athérosensible), Bull.Assoc.Anat., 58:571

Haley, N.J., Shio, H., and Fowler, S., 1977, Characterization of
    lipid-laden aortic cells from cholesterol-fed rabbits.I. Resolu-
    tion of aortic cell populations by Metrizamide density gradient
    centrifugation, Lab.Invest., 37:287
Haust, M.D., 1980, Milieu and function of arterial wall. The clues to
    unique reactivity, in "Atherosclerosis V", A.M.Gotto, L.C. Smith
    and B. Allen,eds., Springer Verlag, New York-Heidelberg-Berlin
Haust, M.D., 1977, Arterial endothelium and its potential, in "Athero-
    sclerosis. Metabolic, morphologic and clinical aspects", Adv. in
    Exp.Med.Biol., G.W. Manning and M.D. Haust,eds., Plenum Press,
    New York-London
Helin, P., Lorenzen, I., Garbasch, C., and Matthiesen, M.E., 1969,
    Relative immunity to atherosclerosis in rabbits during the
    hair-shedding period, J.Arterioscl.Res., 10:359
Hüttner, I. and Peters, H., 1978, Heterogeneity of cell junctions
    in rat aortic endothelium: a freeze fracture study, J.Ultra-
    structure Res., 64:303
Jaffe, E.A., Nachman, R.L., Becker, C.G., and Minick, C.R., 1973,
    Culture of human endothelial cells derived from umbilical veins.
    Identification by morphologic and immunologic criteria, J.Clin.
    Invest., 52:2745
Jellinek, H., 1974, "Arterial lesions and arteriosclerosis", Akad.
    Kiadò, Budapest
Joris, I. and Majno, G., 1974, Cellular break down within the arterial
    wall, Virchows Arch.A Path.Anat.Histol., 364:111
Joris, I., Stetz, E., and Majno, G., 1979, Lymphocytes and monocytes
    in the aortic intima. An electron microscopic study in the rat,
    Atherosclerosis, 34:221
Jorgensen, L., Packham, M.A., Rowsell, H.C., and Mustard, J.P.,1972,
    Deposition of formed elements of blood in the intima and signs
    of intimal injury in the aorta of rabbit, pig and man, Lab.
    Invest., 27:341
Kibria, G., Heath, D., Smith, P., and Biggar, R., 1980, Pulmonary
    endothelial pavement patterns, Thorax , 35:186
Limas, C., Westrum, B.,,and Limas, C.J., 1980, The evolution of vas-
    cular changes in the spontaneously hypertensive rats, Am.J.Path.,
    98:357
Magnani, B. and Coccheri, S., 1960, Sulla patogenesi della ateromasia
    dietetica nel coniglio: significato dei fenomeri di alterata
    permeabilità vascolare, Giorn.Clin.Med., 41:1103

Mason, R.G. and Balis, J.U., 1980, Pathology of the endothelium, in
    "Pathobiology of cell membranes.II.", B.F. Trump and A.U. Arstila,
    eds.,Academic Press, New York-London-Toronto-Sidney-San Francisco
Mc Mahan, C.A., Mc Gill, H.C., and Wigodsky, H.S., 1978, Optimizing
    selection of sites for tissue blocks from arteries for micro-
    scopic evaluation, Exp.Mol.Path., 28:129
Minick, C.R., Stemerman, M.B., and Insull, W.Jr., 1977, Effect of
    regenerated endothelium on lipid accumulation in the arterial
    wall, Proc.Natl.Acad.Sci.(USA), 74:1724
Moares, J.R. and Stastny, P., 1977, A new antigen system expressed
    in human endothelial cells, J.Clin.Invest., 60:449
Nelson, E., Gertz, S.D., Forbes, M.S., Rennels, M.L., Heald, F.P.,
    Kahn, M.A., Farber, T.M., Miller, E., Husain, M.M., and Earl,
    F.L., 1976, Endothelial lesion in the aorta of egg yolk-fed
    miniature swine: a study by scanning and transmission electron
    microscopy, Exp.Mol.Path., 25:208
Olivecrona, T., Bengtsson, G., Hook, M., and Luidohl, U., 1976, Phys-
    iologic implications of the interaction between lipoprotein
    lipase and some sulfated glycosaminoglycans, in "Lipoprotein
    metabolism", H. Greten,ed., Springer Verlag, Berlin
Oxlund, H., Helin, P., and Lorenzen, I., 1979, Seasonal variations
    in the biophysical properties of rabbit aorta and its suscep-
    tibility to atherosclerosis, Atherosclerosis, 32:397
Payling-Wright, H., 1971, Areas of mitosis in aortic endothelium
    of Guinea-pigs, J.Path., 105:65
Pelikan, P. Gimbrone, M.A., and Cotran, R.S., 1979, Distribution and
    movement of anionic cell surface sites in cultured human vascu-
    lar endothelial cells, Atherosclerosis, 32:69
Richardson, M. and Moore, S., 1980, Preparation of large, curved
    biological surfaces for scanning electron microscopy, Artery,
    6:409
Ross, R. and Glomset, J.A., 1976, The pathogenesis of atherosclerosis,
    New Engl.J.Med., 295:369
Ryan, U.S., Ryan, J.W., Whitaker, C., and Chiu, A., 1976, Localiza-
    tion of angiotensin converting enzyme (kinase II), Part 2.
    Immunocytochemistry and immunofuorescence, Tissue and Cell,
    8:125
Scott, R.F., Thomas, W.A., Reiner, J.M., and Florentin, R.A., 1979,
    Population dynamics of arterial cell during atherogenesis. IX.
    Similarity of endothelium cells loss over intimal smooth muscle
    cell masses (cushions) in aortas of swine fed normolipidemic
    diets for 60 days, Exp.Mol.Path., 31:145

Simionescu, M., 1977, The morphologic basis of normal endothelium
     permeability; intercellular pathways, in "Atherosclerosis.
     Metabolic, morphologic and clinical aspects", Adv. in Exp.Med.
     Biol.,G.W. Manning and M.D. Haust,eds., Plenum Press, New York-
     London

Simionescu, M., 1980, The cell membrane of vascular endothelium as
     revealed by freeze-fracture technique, Second International
     Austrian Atherosclerosis Conference, Vienna 7-12 April

Simionescu, M., Simionescu, N., and Palade, G.E., 1976, Segmental
     differentiations of cell junctions in the vascular endothelium.
     Arteries and veins, J.Cell Biol., 68:705

Simionescu, N., 1980, Studies on the biochemistry of cell surface of
     capillary endothelium, Second International Austrian Athero-
     sclerosis Conference, Vienna 7-12 April

Stary, H., 1980, The role of macrophages in progression and regression
     of primate atherosclerosis, Second International Austrian Athero-
     sclerosis Conference, Vienna 7-12 April

Stebhens, W.E., 1960, Focal intimal proliferation in the cerebral
     arteries, Am.J.Path., 36:289

Stein, O., Chajek, T., and Stein, Y., 1976, Ultrastructural localiza-
     tion of Concanavalin A in the perfused rat heart, Lab.Invest.,
     35:103

Stein, O. and Stein, Y., 1976, High density lipoproteins reduce the
     uptake of low density lipoproteins by human endothelial cells
     in culture, Biochem.Biophys.Acta, 431:449

Stout, L.C. and Thorpe, L.W., 1980, Histology of normal aortas in
     non-human primates with emphasis on diffuse intimal thickening
     (DIT), Atherosclerosis, 35:165

Tokes, Z.A. and Sorgente, N., 1976, Cell surface associated and re-
     leased proteolytic activities of bovine aorta endothelial cells,
     Biochem.Biophys.Res.Comm., 73:965

Toledo, O.M.S. and Mourao, P.A.S., 1980, Sulfated glycosaminoglycans
     in normal aortic wall of different mammals, Artery, 6:341

Velican, D. and Velican, C., 1978, Human coronary arteries.II.Branch-
     ing anatomical pattern and arterial wall microarchitecture,
     Acta Anat., 100:258

Velican, D. and Velican, C., 1980, Comparative study on age-related
     changes and atherosclerotic involvement of the coronary arteries
     of male and female subjects up to 40 years old, in press

Vlodavsky, I., Fielding, P.E., Fielding, C.J., and Gospodarowicz, O., 1978, Role of contact inhibition in the regulation of receptor-mediated uptake of low-density lipoprotein in cultured vascular endothelial cells, Proc.Natl.Acad.Sci.(USA), 75:356

Walton, K.W., 1980, Atherogenic factors intrinsic to the artery wall, in "Atherosclerosis V", A.M. Gotto, L.C. Smith and B. Allen,eds., Springer Verlag, New York-Heidelberg-Berlin

Weber, G., 1977, The influence of hypercholesterolemia upon endothelial glycocalyx, in "Atherosclerosis. Metabolic, morphologic and clinical aspects", Adv.Exp.Med.Biol., G.W. Manning and M.D. Haust, eds., Plenum Press, New York-London

Weber, G., Fabbrini, P., Barbaro, A., and Resi, L.,1972, Reattività della Concanavalina A sull'endotelio aortico di coniglio nella norma e nelle prime fasi dell'aterogenesi sperimentale.(Osservazioni preliminari in microscopia elettronica a trasmissione), Boll.Soc.It.Biol.Sper., 48:1009

Weber, G., Fabbrini, P., and Resi, L., 1973, On the presence of a Concanavalin A reactive coat over the endothelial aortic surface and its modifications during early experimental cholesterol atherogenesis in rabbits, Virchows Arch.Abt.A Path.Anat., 359:299

Weber, G., Fabbrini, P., Resi, L. Jones, R., Vesselinovitch, D., and Wissler, R.W., 1977, Regression of atherosclerotic lesions in Rhesus monkey aortas after regression diet. Scanning and transmission electron microscope observations of the endothelium, Atherosclerosis, 26:535

Weber, G., Fabbrini, P., Resi, L., Pierli, C., and Tanganelli, P., 1978, Regeneration of endothelial cells, in "International Conference on Atherosclerosis", L.A. Carlson, R. Paoletti, C.R. Sirtori and G. Weber,eds., Raven Press, New York

Weber, G., Fabbrini, P., Resi, L., Sforza, V., and Tanganelli, P., 1980, "Lesioni" e "disfunzione" delle cellule endoteliali nella aterogenesi sperimentale e nell'ateroregressione, Arch."De Vecchi" in press

Weber, G., Losi, M., Toti, P., and Vatti, R., 1978, Sulla presenza di cellule simil-endoteliali circolanti nel sangue arterioso periferico di conigli a dieta ipercolesterolica, Giorn.Arterioscl., 3:203

Weber, G., Losi, M., Toti, P., and Vatti, R., 1979, Circulating endothelial-like cells in arterial peripheral blood of hypercholesterolemic rabbits, Artery, 5:29

Wissler, R.W., 1980, The artery wall and the pathogenesis of pro-
    gressive atherosclerosis, in "Atherosclerosis V", A.M. Gotto,
    L.C. Smith and B. Allen,eds., Springer Verlag, New York-Heidel-
    berg-Berlin
Wolf, S. and Werthessen, N.T., 1975, "The smooth muscle of the ar-
    tery", Adv.Exp.Med.Biol., Plenum Press, New York-London
Zemplenyi, T., 1968, "Enzyme biochemistry of the arterial wall",
    Lloyd-Luke Ltd., London

METABOLISM OF THE ARTERIAL WALL

Elspeth B. Smith

Department of Chemical Pathology
University of Aberdeen
Foresterhill, Aberdeen AB9 2ZD
Scotland, United Kingdom

In this Advanced Study Institute other speakers will discuss endothelial prostaglandin metabolism, factors controlling smooth muscle and endothelial cell proliferation, and aspects of lipid metabolism. I will confine my discussion to three other areas of arterial wall metabolism - energy production and the consequences of hypoxia, the intimal and medial cellular environment, and collagen synthesis - and will try to show how these may be inter-related in terms of atherogenesis.

ENERGY PRODUCTION IN ARTERIAL WALL

Oxidative Phosphorylation and Glycolysis

A major pathway of energy production in adequately oxygenated tissues involves oxidation of glucose to pyruvate which then enters the Krebs cycle, and may be completely oxidized to CO_2 and water, as shown in outline in Figure 1. In the absence of adequate oxygen instead of entering the oxidative phosphorylation pathway, pyruvate is converted to lactic acid, which accumulates. Anaerobic glycolysis is an inefficient pathway, producing only 2 moles of ATP and 32 kilocalories per molecule of glucose compared with 32 moles of ATP and 417 kilocalories per molecule of glucose oxidized to CO_2 and water (1). In most mamalian tissues (and fermentation systems in which the phenomenum was first described by Pasteur) restoration of adequate oxygen tension stops lactic acid production and the system returns to the oxidative phosphorylation pathway; this is known as the Pasteur effect.

In several early studies on arterial wall metabolism it was reported that oxygen uptake was low, glycolysis was the major

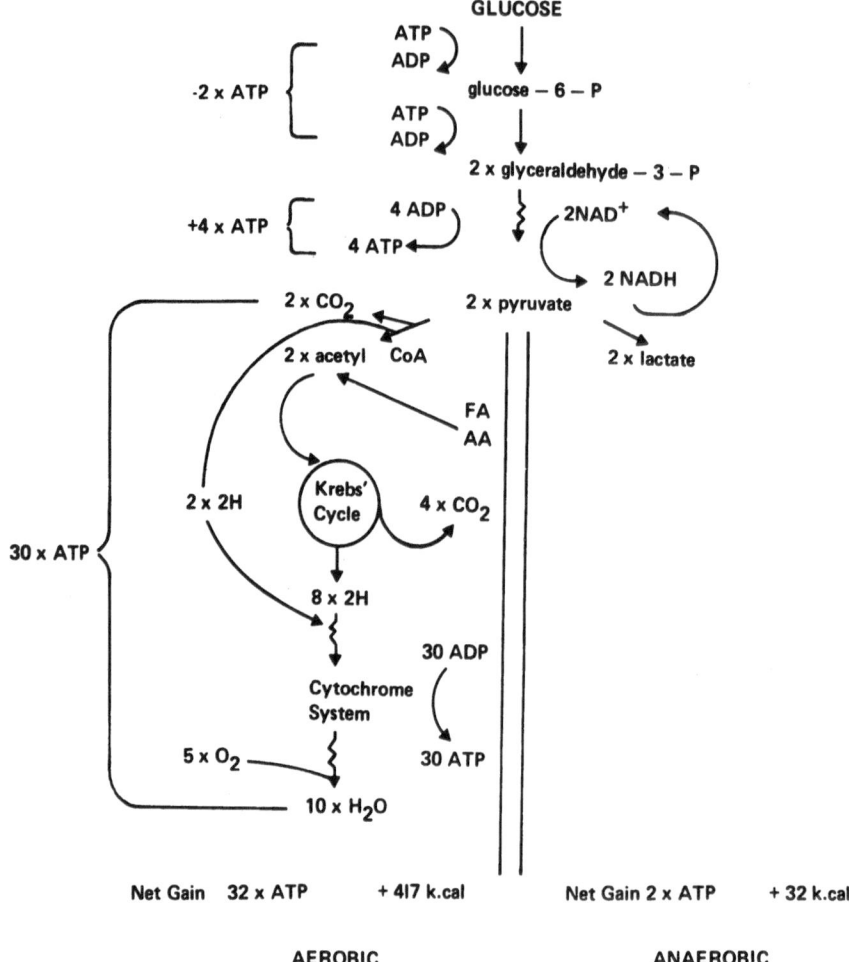

Fig. 1. Pathways of energy production from glucose (adapted
         from ref. 1).

pathway even in the presence of molecular oxygen, and the Pasteur
effect could not be demonstrated or was very weak (2,3,4); the
significance of this so-called "aerobic glycolysis" produced much
speculative discussion.   The arterial wall tissue was studied under
conventional conditions of rapid cooling to 4°C, and sliced or
minced for the metabolic studies.   However, in 1970 Scott et al.
(5) reported that arteries maintained at 37°C following excision
had a much higher oxygen consumption than arteries that were pre-
cooled to 4°C.   Subsequently Morrison and coworkers (6,7) reported
that arteries maintained at 37°C and excised and studied under

conditions in which the integrity of the endothelium was preserved
had a significantly higher oxygen uptake and lower production of
lactic acid than those in which endothelium was disrupted, although
this conclusion has been disputed (8).   With arteries maintained
at 37°C the Pasteur effect has been clearly demonstrated, and it
is enhanced in stretched arteries (7,9).   Thus it appears that
"aerobic glycolysis" was an artefact produced by inappropriate
handling of the tissue.

Oxygen Tension Within the Arterial Wall

If the pattern of energy production changes with oxygen tension
we need to know the oxygen tension within the walls of large arteries
in order to understand their metabolic behaviour.   Several investigators have attempted to make direct measurements by passing oxygen
probes through arteries in experimental animals.   In mid-media,
about 150μm from the endothelial surface and 100μm from the adventitial-medial junction, Niinikoski et al. (10) found a minimum pO_2
of 20mm Hg in aortas of intact rabbits.   Studies on stretched
bovine mesenteric artery suggest that at this level of oxygen significant amounts of lactic acid will be produced (9).   In excised
rabbit aortic arch the pO_2 approached zero at about 100-150μm from
the endothelium (11). It is clear that these measurements are difficult
to perform and subject to artefacts that are discussed by the investigators, but this type of information is essential for understanding
human arterial metabolism, and it is to be hoped that further data
on thick walled arteries and atherosclerotic lesions will be produced in the near future.

A different approach to the problem of arterial wall oxygen
tension was used by Kirk and Laursen (12).   In a classic study they
measured the diffusion coefficients of oxygen, glucose, CO_2 and
lactate across membranes prepared from the intima or media of human
aortas and then used the equation, developed by A.V. Hill for skeletal muscle, to calculate the pO_2 in the wall.   The Hill equation
(13) states that the maximum depth to which oxygen can penetrate
(critical diffusion distance) is a function of the diffusion coefficient, oxygen concentration in the surrounding tissue and rate
of oxygen consumption by the tissue.

$$\text{Critical diffusion distance} = \sqrt{2\ ky/a} \text{ where}$$

k = diffusion coefficient
y = pO_2 of arterial blood
a = rate of oxygen consumption

Using the low values for oxygen consumption obtained with "cold
shocked", disrupted arterial samples Kirk and Laursen obtained a
critical diffusion distance of 900-1000μm, which implies that human
aorta would become anoxic only under the thickest fibrous plaques

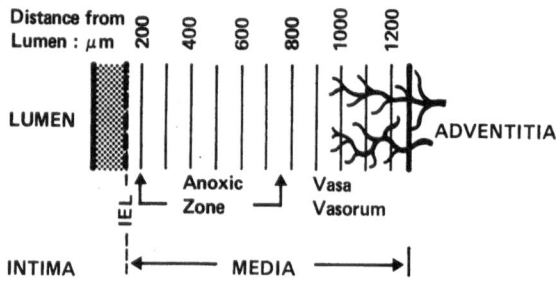

Figure 2.   Spacial relationships in descending thoracic aorta with
             intact IEL: mean values from 9 subjects aged 31-69
             (mean 46.4 years).

and that anoxia would be unlikely to occur in smaller arteries such
as the coronaries.   If, however, the oxygen consumption obtained in
intact vessels kept at 37°C throughout their preparation (6,7) is
used in the equation the calculated critical diffusion distance falls
to 300-350μm.   Back (14) modified the Hill equation and then used
the published data on oxygen tension and diffusion (10,12) to calculate oxygen utilization; the results were comparable with in vitro
findings.   He also showed that there would be reduced penetration
of oxygen at the "back side" of lesions where changes in wall profile
produce flow separation (14).   This is the region of the gelatinous
"tails" of plaques, which seem to be the growing points of the lesions.   Further calculations suggested that the wall will become
anoxic at a depth of 160μm if the blood is uniformly stirred up to
the surface of the endothelium, but if there is a boundary layer with
reduced pO_2, this will reduce the critical diffusion distance to 95μm
(15).   However, using computer modelling Sneiderman et al. (16)
concluded that a boundary layer has little effect, and calculated a
critical diffusion distance of 300μm.

    Thus it appears both from calculations and oxygen probe data
that the arterial wall becomes anoxic at a depth of 150-300μm from
the endothelial surface, and the implications of this in terms of
adult human aorta are illustrated in Figure 2.   In the 30-69 age
group about 40% of samples of apparently normal intima exceed 150μm
in thickness (17), and the total thickness of intima plus media is
about 1200μm (18) with the vasa vasorum extending 300μm into the
media from the adventitial-medial boundary.   Thus it appears that
the anoxic zone starts in the region of the intimal-medial boundary
and extends through the innermost 450-500μm zone of the media.

Accumulation of Lactic Acid

As we have already seen, the metabolic consequence of oxygen deficiency is production of lactic acid. I am not aware of any direct measurements of arterial lactate concentration, but in skeletal muscle Hill (13) showed that the steady state concentration $(y1)$ is related to rate of production and distance from the surface.

$$y^1 = -ab^2/_{2k} + ab^2/_k + y^0$$

where: a = rate of production of lactic acid
b = distance from the blood supply
k = diffusion coefficient in the tissue
y^0 = concentration in the plasma

Using the lactic acid production rate of 24µmol/g wet artery/hour found in stretched anoxic artery (9) and the diffusion coefficients measured by Kirk and Laursen (12) gives calculated lactic acid concentrations of 33mg/100g wet tissue at 300µm from the luminal surface, 51mg/100g at 400µm, which would correspond to the centre of a small plaque, and 174mg/100g at a depth of 800µm, corresponding to the centre of a large plaque. In normal adult aorta the maximum distance from either lumen or vasa vasorum occurs in inner media at about 450µm; here the calculated lactic acid concentration will be 65mg/100g wet tissue. These are crude calculations that ignore effects of bulk flow across the wall. Lactic acid accumulation presumably leads to lowering of tissue pH, the change being dependent on the buffering capacity of the tissue fluid. Albumin concentration in intima and inner media is 25% of plasma concentration (18) and to get an idea of the extent of the pH change we measured the pH of different concentrations of lactic acid in 25% serum. The pH at 50mg/100ml was 6.9, it fell to 6.6 at 65mg/100ml and to 4.8 at 175mg/100ml.

Effect of Hypoxia on Cells

Studies on hypoxia have been made mainly in acute experimental heart infarction, where it produces fall in pH, depletion of cellular glycogen and potassium and reduction in stability of lysosomes (19). Increased leakage of enzymes from the lysosomes occurs within 15 minutes of coronary artery occlusion, and this reduction in latency is particularly marked for cathepsin D (20). Decrease in lysosomal stability was demonstrated biochemically before there was any ultrastructural evidence of cell damage, and it has been suggested that release of acid hydrolases is the primary event that then leads to cell necrosis (21). Hypoxia does not affect the stability of isolated lysosomal preparations (22) and there is evidence that the loss of stability is related to tissue pH; at pH 6 lysosomes remained relatively stable but at pH 5 the proportion of nonsedimentable enzyme activity doubled in 15 minutes (23).

To extrapolate from the acutely infarcted heart to the chronically hypoxic thickened intima may not be justified, but it does
suggest a possible mechanism for the observed changes in early atherosclerotic lesions.   In the centres of large, early gelatinous lesions (400-800μm from the lumen) the architecture of the linear collagen bundles becomes disturbed, producing a network which may be
coated with fine, perifibrous lipid droplets (24).   Low density
lipoprotein (LDL) is rapidly destroyed in incubated intima below
pH 5.5, by an enzyme with the properties of a cathepsin (25);   a
lysosomal cathepsin Bl that degrades native collagen has been
described in rat liver (26), thus leakage of lysosomal enzymes could
account for the first irreversible changes in lesions.   Direct
measurements of arterial wall pH are urgently needed;   the methodology developed for measurement in myocardium should be applicable to
arterial wall, and has recently been reviewed (27).

In cultured cells the binding and internalization of LDL were
not changed by hypoxia, but the rate of degradation was reduced, and
this might lead to lipid accumulation (28).   However, it seems unlikely that hypoxia can be a factor in the development of the fat-
filled cells of fatty streaks.   These usually lie near the surface
of the intima;   in a study of 50 aortas the depth of the centre of
the main band of fat-filled cells in fatty streaks was within 50μm
of the lumen in more than 80% of samples (17).

If hypoxia, lactic acid accumulation and leakage of lysosomal
acid hydrolases are indeed important factors in progression of proliferative lesions into advanced plaques, one must ask why we do not
see medial necrosis?   The calculated pH in the centre of the vascular zone of the human aorta was 6.6, which may not be low enough
significantly to impair lysosomal stability, but another factor may
be the cellular environment.

PLASMA MACROMOLECULES IN INTIMA AND MEDIA

Plasma Proteins in Intima

Using the method of immunoelectrophoresis directly from the
tissue into an antibody-containing gel we have now measured eleven
different plasma proteins in normal intima and lesions;   it seems
probable that all plasma proteins are present.   The concentration of
LDL in normal aortic intima from normotensive subjects is highly
correlated with plasma cholesterol level (29,30), and it is retained to a greater extent than other plasma proteins.   The immunopeaks
can be measured in terms of a standard plasma and, if a blood sample
is available from the patient, in terms of the patient's own serum
or plasma.   From the standard plasma the absolute concentrations in
both intima and the patients' plasma can be calculated.   From the
patients' plasma intimal concentrations can be calculated in terms
of microlitres of plasma from which each antigen was derived.

Table 1.   Plasma Proteins in the Inner Layer of Normal Intima
           (mean thickness 78 μm)

| (n = 15) | M.W. x 10^4 | % of LDL Retention | Concentrations mg/100 cc | |
			Intima	Patient's plasma
LDL	240	100	697	351
α₂-macro	72.5	42.0	159	195
Fibrinogen (n=4)	34.0	32.6	275	552
HDL (apo-Al)	18 ⌉ 35 ⌋	18.4	83	228
Albumin	6.7	15.7	679	2917

If the relation between intimal and plasma concentrations were the same for all proteins they would all be in the same "plasma volume" in a particular tissue sample.   In fact there are large differences between different proteins, and we have called the concentration expressed as microlitres of the patients' plasma "retention".   Retention of LDL in each sample is taken as 100, and the relative retention of other antigens expressed as per cent of LDL retention.

Retention of different plasma proteins is directly proportional to molecular weight (Table 1);   albumin retention is only 16% of LDL retention, and the relation is almost linear, which suggests that there is no specific complex formation by the free LDL fraction (18). This is further supported by comparison of the inner (luminal) and outer layers of intima;   in outer intima the concentration of all plasma proteins decreases to approximately the same extent, suggesting that there is neither specific binding nor molecular sieving within the intima (18).

Plasma Proteins in Media

In samples of aorta in which, by light microscopy, the internal elastic lamina appears to be intact there is an abrupt fall in concentration of plasma proteins immediately outside it, in inner media. Concentrations expressed as per cent of concentration in intima are shown in Table 2;   there is now an inverse relation with molecular weight, with LDL less than 1% and albumin 25-30% of intimal concentration.   This suggests that the IEL is acting as an ultrafilter and that the IEL and not endothelium, is the major barrier to LDL in human aorta (18).

This conclusion has interesting implications for the interpretation of experimental findings (31).   It has long been difficult to reconcile the high concentrations of LDL found in human intima with the very low permeabilities to labelled macromolecules reported in a variety of experimental animals.   In all experimental animals, including young pigs, the intima is only 4-8μm in thickness;   in most

Table 2.   Concentration of Plasma Proteins in Inner
Media from 9 Subjects with Intact IEL

	Per cent of intimal concentration
LDL	0.4
α_2-macroglobulin	12.0
HDL	15.8
Transferrin	19.5
Albumin	26.5

experiments the innermost layer studied was 50-100μm thick, thus it
contained 85-95% media, consequently the measurements mainly reflect
permeability of the IEL, and provide almost no information about
endothelial permeability.   This conclusion is strengthened by the
findings in rabbit aortas in which the mid-medial concentration of
^{125}I-LDL equilibrated at a level of 0.4% of plasma concentration
between 4 and 67 hours (32);   this is comparable to the steady-state
concentration in inner media from human aortas with intact IEL
(Table 3).

Concentrations of plasma proteins in arterial tissue fluid.   If
we calculate intimal and medial concentrations on a volumetric basis
we find the surprising result that the concentration of LDL in intima
is nearly twice the plasma concentration but albumin is only 20% of
plasma concentration (Tables 1 & 3).   Calculation in terms of the
tissue fluid component (18) gives even more bizzare results with
intimal LDL concentration more than thirty times that in lymph
whereas medial LDL is only a quarter of the lymph concentration
(Table 3).

Rate of destruction of LDL is highly dependent on substrate
concentration (25) and the differences in intimal and medial concentrations may account for the gradual accumulation with age of
perifibrous lipid in the deep layers of normal intima whereas there
is much less accumulation in inner media, outside an intact IEL,
despite more profound hypoxia and higher calculated lactic acid
concentrations.

These findings also raise a number of questions with respect to
current concepts of atherogenesis.   For example, studies on LDL uptake in cultured cells are normally made at about 10% serum concentration where the high affinity receptors are already saturated,
but the indigenous normal SM cells are constantly exposed to 200%
LDL and only 25% albumin.   "Endothelial damage" or "increased
endothial permeability" are frequently cited as initiating factors;
permeability is a physico-chemical concept which may not be appropriate for endothelium in which macromolecules appear to be transported across the cells in pinocytotic vesicles (33).   In normal

Table 3.   Plasma Proteins in Intima and Media

| | Per Cent of Plasma Concentration | | | | |
| | INTIMA | | INNER MEDIA | | |
	Whole Tissue	Extracellular Fluid	Whole Tissue	Extracellular Fluid	LYMPH
LDL or apo-B	187	260	0.6	2	8
α_2-macro	77		3.7		
HDL (apo-Al)	37		1.5		<7
Albumin	20	28	7.3	24	32

intima the concentration of LDL is twice the plasma concentration, so damage or removal of the endothelial barrier should allow re-equilibration with plasma, and thus reduction in intimal LDL concentration. Robertson has also concluded that endothelium is not a barrier to passage of blood components to the underlying arterial wall (34).

COLLAGEN SYNTHESIS

Collagen is a major component of normal intima and media as well as fibrous atherosclerotic lesions. Collagen metabolism has been studied in several ways - changes in concentration, which may be misleading if other components are changing at the same time, absolute changes in amount per unit segment, changes in specific activity of hydroxyproline, and changes in activity of the enzyme prolyl hydroxylase. These various methods of study do not always lead to the same conclusions.

Influence of Mechanical Factors and Lactic Acid

A major stimulant of collagen synthesis is mechanical stress or stretching. This has been studied in developing pulmonary and aortic trunks (35), in cultures of SMCs grown on stretched elastin membranes (36), in intermittently stretched diaphragm (37) and in arteries and veins subjected to different degrees of pulsatile flow (38).

In cell cultures collagen synthesis seems to be specifically stimulated by lactic acid (39,40). Thus one can postulate that focal SMC proliferation will produce hypoxic areas with concomitant production of lactic acid, which will be greater in stretched (pulsatile) arteries, and will stimulate collagen production.

Influence of Hypercholesterolaemia

It can be seen in Table 4 that the effects of cholesterol feeding on aortic collagen synthesis are remarkably varied. The variable response may be related to species - thus the

Table 4. Effect of Cholesterol Feeding on Collagen Synthesis:
          Changes Relative to Control Animals

Species	Measurement	Lesion Area	Normal Area	Reference
Dog	Collagen synthesis	+++	No change	41
Pigeon	Collagen synthesis	+++	No change	42
Pig	Prolyl hydroxylase	++	++	43
Pigeon	Prolyl hydroxylase	No change	No change	44

rabbit produces a further variant in which there is stimulation of
collagen synthesis in lesions but this is non-specific and the ratio
collagen synthesis/total protein synthesis is unchanged (45).   It
may also be related to the detailed composition of the diet: peanut
oil is particularly atherogenic, producing low-lipid but highly proliferative, collagen-rich lesions (46) and in rabbits fed peanut oil
with cholesterol there was a specific increase in collagen synthesis
(47).   Most reports suggest that cholesterol feeding does not have
a direct stimulating effect on collagen synthesis; increased collagen
synthesis seems to be a consequence of proliferative plaque development rather than an initiating factor.

CONCLUDING REMARKS

     In this discussion I have tried to suggest how the precarious
oxygen supply in large arteries and the remarkable extracellular
environment in human intima may interact to produce some of the
changes that we see in atherosclerotic lesions.   I have stressed
these two areas because I believe that they are fundamental to our
understanding of what goes wrong in the intima, and in the hopes of
stimulating further research.   We need more and accurate information
on arterial wall oxygen tension, lactic acid concentration and pH,
and quantitative information on the transport of plasma macromolecules across the endothelium.

REFERENCES

1.   A.L. Lehninger, Biochemistry, Worth Publishers, New York (1970).
2.   J.E. Kirk, P.G. Effersoe, and S.P. Chiang, The rate of respirat-
       ion and glycolysis by human and dog aortic tissue, J.Gerontol.,
       9: 10 (1954).
3.   A.L. Lehninger, The metabolism of the arterial wall, in, "The
       Arterial Wall", Lansing, A.I., ed., Williams & Wilkins,
       Baltimore (1959).
4.   A.F. Whereat, Atherosclerosis and metabolic disorder in the
       arterial wall, Exp.Mol.Pathol., 7: 233 (1967).

5. R.F. Scott, E.S. Morrison, and M. Kroms, Effect of cold shock on respiration and glycolysis in swine arterial tissue, Am.J. Physiol., 219: 1363 (1970).

6. A.D. Morrison, L. Berwick, L. Orci, and A.I. Winegrad, Morphology and metabolism of an aortic intima-media preparation in which intact endothelium is preserved, J.Clin.Invest., 57: 650 (1976).

7. A.D. Morrison, L. Orci, L. Berwick, A. Perrelet and A.I. Winegrad, The status of the arterial endothelium in experimental studies, Atherosclerosis Revs., 3: 125 (1978).

8. E.S. Morrison, J. Frick and M. Kroms, The effects of O_2 concentration and albumin on respiration and aerobic glycolysis in rabbit aortic intima-media, Biochem.Med., 20: 279 (1978).

9. H.J. Arnqvist and L. Lundholm, Influence of oxygen tension on the metabolism of vascular smooth muscle: demonstration of a Pasteur effect, Atherosclerosis, 25: 245 (1976).

10. V. Niinikoski, C. Heughan and T.K. Hunt, Oxygen tensions in the aortic wall of normal rabbits, Atherosclerosis, 17: 353 (1973).

11. E.R. Jurrus and H.S. Weiss, In vitro tissue oxygen tensions in the rabbit aortic arch, Atherosclerosis, 28: 223 (1977).

12. J.E. Kirk and T.J.S. Laursen, Diffusion coefficients of various solutes for human aortic tissue, with special reference to variation in tissue permeability with age, J.Gerontol., 10: 288 (1955).

13. A.V. Hill, The diffusion of oxygen and lactic acid through tissues, Proc.Roy.Soc. (London) Series B, 104: 39 (1928-29).

14. L.H. Back, Analysis of oxygen transport in the avascular region of arteries, Math.Biosci., 31: 285 (1976).

15. L.H. Back, D.W. Crawford, personal communication.

16. G. Sneiderman and T.K. Goldstick, Significance of luminal plasma layer resistance in arterial wall oxygen supply, Atherosclerosis, 31: 11 (1978).

17. E.B. Smith, Metabolic activities in the arterial wall, in, "Dynamics of Arterial Flow", S. Wolf and N.T. Werthessen, eds., Advan.Exp.Med.Biol., 115: 245 (1979).

18. E.B. Smith and E.M. Staples, Distribution of plasma proteins across the human aortic wall: barrier functions of endothelium and internal elastic lamina, Atherosclerosis, 37: 579 (1980).

19. M.A. Ricciutti, Myocardial lysosome stability in the early stages of acute ischaemic injury, Amer.J.Cardiol., 30: 492 (1972).

20. K. Wildenthal, R.S. Decker, R. Poole, E.E. Griffin and J.T. Dingle, Sequential lysosomal alterations during cardiac ischaemia: 1. Biochemical and immunohistochemical changes, Lab.Invest., 38:656 (1978).

21. K. Wildenthal, Lysosomal alterations in ischaemic myocardium: result or cause of myocardial damage?, J.Mol.Cell.Cardiol., 10: 595 (1978).

22.   C. DeDuve and H. Beaufay, Tissue fractionation studies: 10.
      Influence of ischaemia on the state of some bound enzymes
      in rat liver, Biochem.J., 73: 610 (1959).
23.   L. Gordis and H.M. Nitowsky, Lysosomes in human cell cultures:
      kinetics of enzyme release from injured particles, Exp.Cell
      Res., 38: 556 (1965).
24.   E.B. Smith and R.H. Smith, Early changes in aortic intima,
      Atherosclerosis Rev., 1: 119 (1976).
25.   E.B. Smith and I.B. Massie, Destruction of endogenous low
      density lipoprotein in incubated intima, Atherosclerosis,
      26: 427 (1977).
26.   M.C. Burleigh, A.J. Barrett and G.S. Lazarus, Cathepsin B_1: a
      lysosomal enzyme that degrades native collagen, Biochem.J.
      137: 387 (1974).
27.   P.A. Poole-Wilson, Measurements of myocardial intracellular pH
      in pathological states, J.Mol.Cell.Cardiol., 10: 511 (1978).
28.   J.J. Albers and E.L. Bierman, The effect of hypoxia on uptake
      and degradation of low density lipoproteins by cultured
      human arterial smooth muscle cells, Biochim.Biophys.Acta,
      424: 422 (1976).
29.   E.B. Smith and R.S. Slater, Relationship between low density
      lipoprotein in aortic intima and serum lipid levels, Lancet
      i, 463 (1972).
30.   E.B. Smith, The relationship between plasma and tissue lipids in
      human atherosclerosis, Advan.Lipid Res., 12: 1 (1974).
31.   E.B. Smith, Biochemical studies on permeability and the inter-
      action between blood constituents and arterial components
      in atherosclerosis, in, "Atherosclerosis V", A.M. Gotto,
      L.C. Smith and B. Allen, eds., Springer-Verlag, Berlin, 121
      (1980).
32.   R.L. Bratzler, G.M. Chisholm, C.K. Colton, K.A. Smith and R.S.
      Lees, The distribution of labelled low density lipoproteins
      across the rabbit thoracic aorta in vivo, Atherosclerosis,
      28: 289 (1977).
33.   G. Thorgeirsson and A.L. Robertson, The vascular endothelium -
      pathobiological significance, Amer.J.Path., 93: 803 (1978).
34.   A.L. Robertson, Arterial endothelium in the initial stages of
      atherogenesis, in, "Atherosclerosis V", A.M. Gotto, L.C.
      Smith and B. Allen, eds., Springer-Verlag, Berlin 103 (1980).
35.   D.Y.M. Leung, S. Glagov, J.M. Clark and M.B. Mathews, Mechan-
      ical influences on the biosynthesis of extracellular macro-
      molecules by aortic cells, in, "Extracellular Matrix Influ-
      ences on Gene Expression", H.C. Slavkin and R.C. Greulich,
      eds., Academic Press, New York, 633 (1975).
36.   D.Y.M. Leung, S. Glagov and M.B. Mathews, Cyclic stretching
      stimulates synthesis of matrix components by arterial smooth
      muscle cells in vitro, Science, 191: 475 (1976).
37.   P.J. Reeds, R.M. Palmer and R.H. Smith, Protein and collagen
      synthesis in rat diaphragm muscle incubated in vitro: the
      effect of alterations in tension produced by electrical or
      mechanical means, Int.J.Biochem., 11: 7 (1980).

38.   G.M. Fischer, M.L. Swain and K. Cherian, Increased vascular
      collagen and elastin synthesis in experimental atheroscler-
      osis in the rabbit, Atherosclerosis, 35: 11 (1980).
39.   U. Langness and S. Udenfriend, Collagen proline hydroxylase
      activity and anaerobic metabolism, in, "Biology of the
      Fibroblast", E. Kulonen and J. Pikkarainen, eds., Academic
      Press, New York, 373 (1973).
40.   R. Schwarz, L. Colarusso and P. Doty, Maintenance of differ-
      entiation in primary cultures of avian tendon cells, Exp.
      Cell Res., 102: 63 (1976).
41.   K.G. McCullagh and L.A. Ehrhart, Increased arterial collagen
      synthesis in experimental canine atherosclerosis, Athero-
      sclerosis, 19: 13 (1974).
42.   K.G. McCullagh and L.A. Ehrhart, Enhanced synthesis and
      accumulation of collagen in cholesterol-aggravated pigeon
      atherosclerosis, Atherosclerosis, 26: 341 (1977).
43.   G.C. Fuller, E. Miller, T. Farber and E. Vanloon, Aortic
      connective tissue changes in miniature pigs fed a lipid-rich
      diet, Connective Tiss.Res., 1: 217 (1972).
44.   R.W. St. Clair, J.J. Toma and H.B. Lofland, Proline hydroxylase
      activity and collagen content of pigeon aortas with natur-
      ally-occurring and cholesterol-aggravated atherosclerosis,
      Atherosclerosis, 21: 155 (1975).
45.   L.A. Ehrhart and D. Holderbaum, Stimulation of aortic protein
      synthesis in experimental rabbit atherosclerosis, Athero-
      sclerosis, 27: 477 (1977).
46.   D. Vesselinovitch, G.S. Getz, R.H. Hughes and R.W. Wissler,
      Atherosclerosis in the Rhesus monkey fed three food fats,
      Atherosclerosis, 20: 303 (1974).
47.   L.A. Ehrhart and D. Holderbaum, Aortic collagen, elastin and
      non-fibrous protein synthesis in rabbits fed cholesterol
      and peanut oil, Atherosclerosis, 37: 423 (1980).

AORTIC LIPID METABOLISM

David Kritchevsky

The Wistar Institute of Anatomy and Biology
36th and Spruce Streets
Philadelphia, Pennsylvania 19104, U.S.A.

Windaus (1) was one of the first investigators to demonstrate the presence of inordinate amounts of cholesterol ester in atherosclerotic aortas. Analyses of the lipids of aortic plaques revealed that their composition was roughly comparable to that of plasma (Table 1) and these findings lent support to the hypothesis that aortic lipids were derived from the blood. However it has been shown that aortas of children contain either very little cholesteryl ester or none (6), so that aortic lipids cannot be due to filtration of plasma. The aorta can synthesize fatty acids (7) and phospholipids (8, 9) but its cholesterol is derived almost entirely from the blood (10, 11). Calculations of the extent of transfer of cholesterol into the human aorta based on the formula: that transfer (mg. cholesterol per day per gm tissue) is the product of aortic cholesterol content (mg/g) by aortic cholesterol specific activity (% plasma) divided by days times 100 suggest a net transfer of 0.30-0.45 mg/gm aorta/day (12-14).

Comparison of the lipids of young and old human aortas (15) grouped by age and severity of atherosclerotic lesions reveals that as percentage of composition, cholesteryl ester increases

*Supported, in part, by grants HL-03299 and HL-23625 and a Research Career Award (HL-0734) from the National Institutes of Health.

by a factor of 9, while cholesterol content doubles and that of
the other lipids decreases. Calculated as percentage of lipid
within the dry aorta and going from no lesions (avg age 6 yrs) to
class III lesions (avg age 58 yrs) the increase in each lipid
class is: triglyceride, 2.5-fold; free fatty acids, one-third
decrease; phospholipid, 2.5-fold; free cholesterol, 10.3-fold and
esterified cholesterol, 58-fold.

The data suggest local synthesis of esterified cholesterol.
Newman and Zilversmit (11) have shown that the cholesteryl ester
content of aortas of cholesterol-fed rabbits increases to a much
greater extent than does the content of free cholesterol. The
ratio of aortic free to esterified cholesterol is 10-20 in normal
rabbits and 0.4-2.0 in atherosclerotic ones (16-18).

Smith (19) analyzed atheromatous areas of human aortas and the
adjacent normal tissue and found that the ratio of free to esterified cholesterol was considerably higher in the normal tissue. The
free/ester cholesterol ratio was 0.80 in the normal aorta and 0.37
in the fatty streak. The ratio of cholesteryl ester linoleic to
oleic acid was 1.14 in the normal aorta and 0.39 in the fatty
streak; the ratio in serum was 1.57. These data suggest a fatty
acid specificity in the esterification of aortic free cholesterol.
Dayton and Hashimoto (20) reviewed data on cholesterol metabolism
in aorta and concluded that about half of the cholesteryl ester
was derived from the blood. The cholesterol moiety of cholesteryl
ester can leave a cell only after hydrolysis (21, 22) thus the
net movement of cholesterol into and out of aorta would require
both synthesis and hydrolysis of aortic cholesterol. We have
studied an aortic enzyme preparation which, under proper conditions
can synthesize or hydrolyze cholesterol (23, 24). The enzyme(s)
resembles pancreatic cholesterol esterase (21) insofar as it is
present in acetone powder of the tissue and requires no activation.

We have found fatty acid specificities for synthesis (oleic
acid) and hydrolysis (linoleic acid). This would explain further
the accumulation of cholesteryl oleate in aorta. Synthesis (of
oleate) proceeds most efficiently with cholesterol or cholestanol.

A study of the ratios of cholesteryl ester synthesis to
hydrolysis (S/H) in aortas of a number of species reveals that the
ratio is higher in aortas of susceptible species (Table 3) (25);
of special interest in this regard are our studies of S/H ratios
in pigeon aortas. The White Carneau pigeon exhibits spontaneous
atherosclerosis, principally in the distal portion of the aorta.
The Show Racer pigeon is free of atherosclerosis. The White
Carneau aortic S/H ratio is higher than that of the Show Racer.
The ratio of activities (WC/SR) in the proximal portion of the

aorta is 1.35 and in the distal portion it is 1.83 (26) (Table 4).

We have also studied the effects of drugs on aortic S/H ratios. When rabbits were fed cholesterol their aortic S/H rose by 180%. When the diet also contained a hypolipidemic drug (nicotinic acid, clofibrate, β-sitosterol, D-thyroxine) the ratio reverted towards normal (27) (Table 5). In another experiment we fed two groups of rabbits an atherogenic diet for 30 days with the diet of one group also containing 0.5% nicotinic acid. The purpose of this study was to investigate the effect of a duration of feeding which would produce atherosclerosis. The results are shown in Table 6 (28). Average atherosclerosis, graded on a 0-4 basis (29), was reduced by 65% by the drug and the S/H ratio which had risen three-fold upon cholesterol feeding fell to within the normal range. We have also studied the effects of two new hypolipidemic drugs, lipanthyl and pyrinixil, in this system. Both drugs reduce S/H ratios in normal rabbits but only lipanthyl does so in cholesterol-fed rabbits (Table 7) (30). Compactin (5 mg/kg p.o.) reduces the S/H ratio in normal rabbit aorta from 3.99 to 1.70.

The aortic enzyme system which accounts for cholesteryl ester synthesis and hydrolysis offers a very useful tool to study the effects of various dietary or pharmaceutical regimens.

Table 1. Lipids of Human Plasma and Aortic Plaques (% of Total Lipids)

Lipid	Plaques (3)*	Plasma (2)*
Cholesterol		
Total	56.8	50.2
Free	19.2	13.1
Free/Ester	0.51	0.35
Triglyceride	27.3	22.4
Phospholipid	15.2	26.7

*Averaged number of studies (Refs. 2-4, plaques; 2, 5, plasma).

Table 2. Lipids of Human Aortas*

	Stage of Atherosclerosis			
	0	I	II	III
No.	9	6	9	9
Avg., Age, yrs.	6	29	58	56
Lipids, % dry wet	3.6	3.9	10.9	16.0
Lipids, % Composition				
Triglycerides	15.6	14.3	11.7	9.5
Free Fatty Acids	9.6	6.4	2.7	1.2
Phospholipid	60.9	55.0	33.2	33.7
Cholesterol, free	8.1	12.0	19.1	19.7
Cholesterol ester	4.1	13.4	36.0	36.2
FC/EC	1.98	0.90	0.53	0.54
Lipids, % in Dry Aorta				
Triglycerides				
Free Fatty Acid	0.3	0.2	0.3	0.2
Phospholipid	2.2	2.1	3.6	5.4
Cholesterol, free	0.3	0.5	2.1	3.1
Cholesterol ester	0.1	0.5	3.9	5.8
FC/EC	3.00	1.00	0.54	0.53

*After Bottcher (15).

Table 3. Ratio of Cholesteryl Ester Synthetase
to Hydrolase (S/H) in Aortas of Various
Species*

Species	S/H
Man	0.88
Baboon	0.85
Pig	1.06
Rabbit	0.95
Chicken	0.98
Rat	0.58
Mouse	0.62
Dog	0.32

*Kritchevsky and Kothari (25).

Table 4. Cholesteryl Esterase Activity in Pigeon Aorta (White Carneau vs. Show Racer)*

Aorta Portion	S/H Ratio ± S.E.M.	
	White Carneau	Show Racer
Proximal	0.65 ± 0.08^a	0.48 ± 0.03^b
Distal	0.95 ± 0.10	0.52 ± 0.05^b

*Kritchevsky and Kothari (26).
[a]vs. WC distal, $p < 0.05$
[b]vs. WC distal, $p < 0.01$

Table 5. Influence of Hypolipidemic Drugs on Aortic Cholesterol Esterase of Rabbits (4 Rabbits per group: Fed 10 Days)*

Regimen	Serum Cholesterol[†]	Aortic S/H[§]
Basal (B)	76 ± 7	1.00
B + 5% Corn Oil (BC)	83 ± 8	1.64
BC + 1% Cholesterol (BCC)	463 ± 36	2.82
BCC + 0.5% Nicotinic Acid	285 ± 54	0.91
BCC + 0.3% Clofibrate	464 ± 87	2.34
BCC + 1% β Sitosterol	366 ± 98	1.57
BCC + D-Thyroxine (0.5 mg/day)	279 ± 55	1.88

*Kritchevsky et al. (27)
[†]mg/dl ± S.E.M.
[§]Values normalized to basal diet.

Table 6. Influence of Nicotinic Acid (0.5%) Administration on Aortic Metabolism of Rabbits Fed an Atherogenic Diet* (4 Rabbits per Group; Fed 30 Days)

Diet[f]	Cholesterol		Atherosclerosis[#]	S[a]	H[a]	S/H
	Serum (mg/dl)	Liver (mg/100 g)				
B	48 ± 3	171 ± 57	0.00	10.1 ± 1.1	24.1 ± 6.1	0.41
BC	353 ± 87	977 ± 435	0.85	22.5 ± 1.8	16.5 ± 1.1	1.36
BCN	391 ± 85	558 ± 204	0.30	12.3 ± 0.4	21.6 ± 3.5	0.57

*Hirsch and Kritchevsky (28).
[f]B, laboratory ration; BC, B plus 5% corn oil and 1% cholesterol; BCN, BC plus 0.5% nicotinic acid.
[#](Arch plus Thoracic) ÷ 2 (Graded on 2–4 Scale).
[a]S, Synthesis, nmoles cholesterol esterified/mg P/hr.
[a]H, Hydrolysis, nmoles cholesterol liberated/mg P/hr.

Table 7. Influence of Lipanthyl and Pirinixil on Aortic Cholesterol Esterase [EC 3.1.1.13] in Rabbits* (4 Rabbits per Group; Fed 22 Days)

Group[+]	Average Atherosclerosis[¶]	Esterase Activity[§]		S/H
		S	H	
B	0	168 ± 6	12 ± 1	14.0
BL	0.13	173 ± 8	21 ± 3	8.2
BP	0	143 ± 3	21 ± 4	6.8
BC	0.32	171 ± 6	27 ± 2	6.3
BCL	0.19	174 ± 6	39 ± 1	4.5
BCP	0.38	168 ± 3	24 ± 3	7.0

*Kritchevsky and Singer (30).
[+]B, laboratory ration; BL, B plus .15 mg lipanthyl/day; BP, B plus 10 mg Pirinixil/day p.o.; BC, B plus 5% corn oil and 1% cholesterol; BCL, BC plus lipanthyl; BCR, BC plus pirinixil.
[¶](Arch plus Thoracic)÷2.
[§]S, nmoles cholesterol esterified/mg protein/hour.
H, nmoles cholesterol liberated mg protein/hour.

REFERENCES

1. A. Windaus, Über den Gehalt normaler und Atheromatöser Aorten an Cholesterin und Cholesterinestern, Hoppe-Seyler Z. Physiol. Chem. 67:174 (1910).
2. S. Weinhouse and E.F. Hirsch, Chemistry of Atherosclerosis. I. Lipid and calcium content of the intima and of the aorta with and without atherosclerosis, Arch. Path. 29:31 (1940).
3. I. H. Page, Some aspects of the nature of the chemical changes occurring in atheromatosis, Ann. Int. Med. 14:1741 (1941).
4. J. F. Mead and M. L. Gouze, Alterations in aorta lipids with advancing atherosclerosis, Proc. Soc. Exp. Biol. Med. 106:4 (1961).
5. I. H. Page, E. Kirk and D.D. Van Slyke, Plasma lipids and essential hypertension, J. Clin. Invest. 15:109 (1936).

6. R. F. Scott, R. A. Florentin, A. S. Daoud, E. S. Morrison,
    R. M. Jones and M. S. R. Hutt, Coronary arteries of
    children and young adults. A comparison of lipids and
    anatomic features in New Yorkers and East Africans,
    Exp. Molec. Path. 5:12 (1966).

7. A. F. Whereat, Lipid biosynthesis in aortic intima from
    normal and cholesterol-fed rabbits, J. Atheroscler.
    Res. 4:272 (1964).

8. M. L. Shore, D. B. Zilversmit and R. F. Ackerman, Plasma
    phospholipide deposition and aortic phospholipide
    synthesis in experimental atherosclerosis, Am. J. Physiol.
    181:527 (1955).

9. D. B. Zilversmit and E. L. McCandless, Independence of
    arterial phopholipid synthesis from alterations in blood
    lipids, J. Lipid Res. 1:118 (1959).

10. M. W. Biggs and D. Kritchevsky, Observations with radio-
    active hydrogen (H^3) in experimental atherosclerosis,
    Circulation 4:34 (1951).

11. H. A. I. Newman and D. B. Zilversmit, Accumulation of lipid
    and nonlipid constituents in rabbit atheroma, J. Athero-
    scler. Res. 4:261 (1964).

12. M. W. Biggs, D. Kritchevsky, D. Colman, J. W. Gofman, H. B.
    Jones, F. T. Lindgren, G. Hyde and T. P. Lyon, Observa-
    tions on the fate of ingested cholesterol in man,
    Circulation 6:359 (1952).

13. H. Field, Jr., L. Swell, P. E. Schools, Jr. and C. R.
    Treadwell, Dynamic aspects of cholesterol metabolism
    in different areas of the aorta and other tissues in man
    and their relationship to atherosclerosis, Circulation
    22:547 (1960).

14. R. G. Gould, R. W. Wissler and R. J. Jones, The dynamics
    of lipid deposition in arteries, in: "Evolution of the
    Atherosclerotic Plaque," R. J. Jones, ed., Univ. of
    Chicago Press, Chicago (1963), p. 205.

15. C. J. F. Bottcher, Lipids of the human arterial wall, in:
    "Drugs Affecting Lipid Metabolism," S. Garattini and
    R. Paoletti, eds., Elsevier Publ. Co., Amsterdam (1961),
    p. 54.

16. F. Parker and G. F. Odland, A correlative histochemical,
    biochemical and electron microscopic study of experimen-
    tal atherosclerosis in rabbit aorta with special refe-
    rence to the myo-intimal cell, Am. J. Path. 48:197 (1966).

17. L. Swell, M. D. Law and C. R. Treadwell, Tissue cholesterol
    ester and triglyceride fatty acid composition of rabbits
    fed cholesterol diets high and low in linoleic acid,
    J. Nutr. 76:429 (1962).

18. D. Kritchevsky and S. A. Tepper, Experimental atherosclerosis
    in rabbits fed cholesterol-free diets: Influence of chow
    components, J. Atheroscler. Res. 8:357 (1968).

19. E. B. Smith, The influence of age and atherosclerosis on the chemistry of aortic intima. Part 1 (The lipids), J. Atheroscler. Res. 5:224 (1965).

20. S. Dayton and S. Hashimoto, Origin of fatty acids in lipids experimental rabbit atheroma, J. Atheroscler. Res. 8:555 (1968).

21. J. M. Bailey, Lipid metabolism in cultured cells. III. Cholesterol excretion process, Am. J. Physiol. 207:1221 (1964).

22. G. H. Rothblat and D. Kritchevsky, The excretion of free and ester cholesterol by tissue culture cells: Studies with L5178Y and L-cells, Biochim. Biophys. Acta 144:423 (1967).

23. H. V. Kothari, B. F. Miller and D. Kritchevsky, Aortic cholesterol esterase: Characteristics of normal rat and rabbit enzyme, Biochim. Biophys. Acta 296:446 (1973).

24. H. V. Kothari and D. Kritchevsky, Purification and properties of aortic cholesteryl ester hydrolase, Lipids 10:322 (1975).

25. D. Kritchevsky and H. V. Kothari, Aortic cholesterol esterase in species resistant or susceptible to atherosclerosis, Steroids, Lipids Res. 5:23 (1974).

26. D. Kritchevsky and H. V. Kothari, Aortic cholesterol esterase: Studies in White Carneau and Show Racer Pigeons, Biochim. Biophys. Acta 326:489 (1973).

27. D. Kritchevsky, S. A. Tepper and H. V. Kothari, Effect of hypocholesteremic drugs on aortic cholesterol esterase in cholesterol-fed rabbits, Artery 1:437 (1975).

28. C. Z. Hirsch and D. Kritchevsky, unpublished.

29. G. L. Duff and G. C. McMillan, The effect of alloxan diabetes on experimental cholesterol atherosclerosis in the rabbit, J. Exp. Med. 89:611 (1949).

FROM THE FATTY STREAK TO THE CALCIFIED LESION

Elspeth B. Smith

Department of Chemical Pathology
University of Aberdeen
Foresterhill, Aberdeen AB9 2ZD
Scotland, United Kingdom

INTRODUCTION

Juvenile-type fatty streaks are the earliest lesions that can be recognized by macroscopic inspection of aortas of children. They characteristically appear as small yellow/white dots most frequently in longitudinal lines between the intercostal branches, they stain brilliantly red with macroscopic sudan staining, and Holman reported that they were already present in all children aged more than 3, increased in area rapidly between ages 8-15, and reached a maximum at about age 20 (1).

Microscopically they are characterized by groups of smooth muscle cells (SMCs) that are filled with large lipid droplets; the fat-filled SMCs may be scattered, or in confluent strands or bunches, and may lie directly under the endothelium with no associated intimal thickening, or form part of small proliferative lesions which contain very little extracellular lipid. For many years it was widely believed that they were the precursors of fibrous plaques, and it was postulated that the fat-filled cells disintegrated, releasing sclerogenic lipid that stimulated proliferation of SMCs and collagen. However, there is now evidence from many different sources that suggests that fatty streaks and fibrous plaques develop by separate and independent pathways (2,3).

EARLY LESIONS

Coronary Lesions in Young People

Velican and Velican have recently completed a remarkable systematic microscopical study of the coronary arteries of successive

45

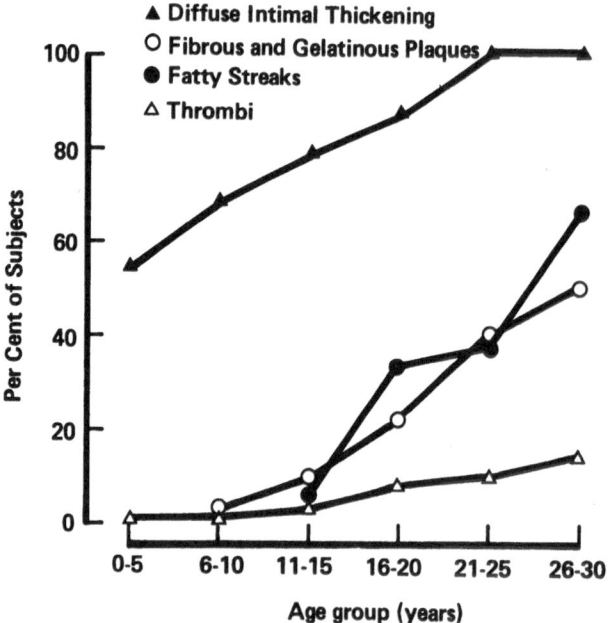

Fig. 1.   Prevalence of coronary artery lesions in different age
           groups (from references 4-7).

age groups from neonates up to age 40, and by recording the frequency
with which different types of intimal change were seen at different
ages they have tried to build up a picture of the course of lesion
development (4,5,6,7).   Some of the main features of their findings
are summarized in Figure 1.   Branch pads were ubiquitous even in
neonates, and diffuse intimal thickening was already present in more
than half the children in the youngest age group, increasing steadily
so that it was present in all subjects over age 20.   Microthrombi,
consisting of well circumscribed accumulations of fibrin and platelets, partially incorporated or on the endothelial surface, were
present in 2-3% of all children up to age 15, and then increased
rather rapidly.   Fibrous and gelatinous lesions (independent of
branching pads) first appeared in the 6-10 age group, whereas fatty
streaks were not detected until ages 11-15.   The incidence of fibrous
and gelatinous plaques and fatty streaks then increased rapidly and
more or less in parallel, a pattern that does not suggest a precursor-
product relationship.

The Velicans record many significant morphological observations.
They did not observe conversion of fatty streaks into atherosclerotic
plaques and concluded that the two types of lesion developed as unrelated pathological processes.   "Advanced" fatty streaks exhibiting

cell disintegration and accumulation of extracellular lipid were
first encountered in the 26-30 age group and increased fairly
rapidly over the next decade, but again they did not observe "further
transitional stages between advanced fatty streaks and atherosclerotic
plaques". Plaques were particularly prone to develop in areas of
diffuse intimal thickening and their early manifestations seemed to
be histolysis, nodular proliferation of smooth muscle cells and insudation. In the third decade lipid became abundant in the plaques
in the form of foam cells which were particularly associated with
areas of insudation, and small pools of extracellular lipid; there
was also "progressive involvement of microthrombi in the early steps
of plaque formation".

Initiating Factors

Studies on the relation in different population groups between
fatty streaking in young people and extent and severity of raised
lesions in older age groups (2), biochemical comparisons of the fatty
acid composition of the cholesterol esters that accumulate intra-
cellularly in SMCs in fatty streaks and extracellularly in developing
plaques (3) and the age-sequence studies in a homogeneous population
described above (4-7) all indicate that fatty streaks and fibrous
plaques develop by independent pathways, but we still do not understand the initiating mechanisms.

Fatty streaks. The fat filled SMCs contain 40-50 mg total lipid per
100 mg lipid-extracted dry tissue, and 60-70% of the lipid is cholesterol ester. This cholesterol ester contains about 50% cholesterol oleate and 12% cholesterol linoleate, a composition that is in
striking contrast to the 50% cholesterol linoleate and 24% cholesterol oleate of the plasma low density lipoprotein (LDL) that is present
in normal intima in high concentration (3,8). Specific accumulation
of cholesterol oleate suggests that the cells are behaving like SMCs
or fibroblasts in culture by binding and internalizing LDL followed
by degradation of the apoprotein, hydrolysis of cholesterol ester
and re-esterification with oleate via the acyl-CoA: cholesterol
acyl transferase pathway (9). However, we do not know why individual cells or groups of cells accumulate cholesterol oleate while
neighbouring cells do not; this is a fascinating problem, but probably not central to the understanding of atherogenesis. In the
immediate vicinity of the fat-filled SMCs the concentration of LDL
is about a quarter of the concentration in adjacent normal intima,
but there is no significant change in the concentrations of albumin
or fibrinogen (3). This suggests that uptake and degradation of
LDL has been greatly accelerated, so that rate of production of
cholesterol oleate exceeds its possible rate of removal. Cultured
SMCs and fibroblasts do not overload themselves when exposed to
normal lipoproteins but cationized LDL leads to cholesterol ester
accumulation (10,11); it might be postulated that these cells are
reacting to modified LDL molecules, but this still does not explain
their localization.

Proliferative lesions.   Early lesions that are not lipid-rich have
been widely overlooked because they are inconspicuous in fresh arteries and do not stain macroscopically with Sudan dyes.   Velican and
Velican (4-7) could not detect these lesions on macroscopic examination below the age of 20, and in young adults one third of the lesions
found microscopically escaped macroscopic detection.

The Velicans differentiate between gelatinous lesions and mucoid
plaques but it seems probable that both these categories are included
in the term gelatinous lesion as it is used by Haust (12) and by the
present author (13,14).   Haust (12) gives the most detailed descrip-
tion of their morphology as revealed by light- and electronmicroscopy
and immunofluorescence.   The salient features are focal oedema and
separation of the connective tissue elements by plasma insudate, reduction in acid glycosaminoglycan (GAG) observed both by staining
(12,6) and by chemical analysis (15), swelling of collagen fibres,
and in larger lesions, strands of fibrin in the insudate.   Fine,
extracellular lipid droplets can be seen on the collagen fibres of
lesions that macroscopically appear yellowish, but are absent in the
greyish lesions.   Within this framework the lesions present a con-
tinuous spectrum from those in which there is predominantly insudation
and separation of SMCs and collagen fibres to relatively compact
masses of proliferated SMCs.   The morphological features are closely
paralleled by the biochemical findings (3,13,14,15).

Again, we do not know which of these changes occurs first, and
what initiates it.   If it is postulated that oedema and increased
insudation of plasma LDL as a result of "endothelial damage" is the
first change we have to remember that normal intima already contains
more than twice the plasma concentration of LDL (8).   Furthermore,
the normal intima of a hypercholesterolaemic, hypertensive subject
may contain twice as much LDL as a gelatinous lesion in a normotensive subject with low serum cholesterol levels (16).   Clearly high
concentration of intimal LDL do not necessarily induce focal SMC
proliferation.   In the younger age groups microthrombi apparently
were not associated with gelatinous or "mucoid" lesions although they
were found on diffusely thickened intima;   only above age 30 were a
small proportion of microthrombi associated with developing lesions
(7).   This does not support the idea that the lesion develops as a
result of stimulation of SMC proliferation initiated by growth factor
released from platelet microthrombi (17).

In the coronary arteries the lesions invariably started in areas
of thickened intima and the sequence of events appeared to be first
"histolysis, followed by nodular proliferations of SMCs, new formation of collagen (but not elastic) fibres, and insudation ...." (7).
This suggests that lysosomal damage, leading to leakage of acid
hydrolases may be the initiating injury;   the areas of intima in
which the lesions start to develop seem to be severely thickened -

150-200μm, which may be close to the critical diffusion distance for oxygen (8), but this does not explain the SMC proliferation.   In the Velicans' data all adolescents (16-20 age group) with "mucoid" plaques smoked 5-12 cigarettes daily (5).   This raises the possibility that a mutagenic substance in tobacco smoke, perhaps carried by LDL, causes cell transformation as suggested by Benditt (18).   Again we are faced with the problem of localization when we know that there is such a high concentration of LDL in normal intima.   However, we know nothing about the cell microenvironment or compartmentalization in intima;   in normal intima the LDL concentration in the immediate vicinity of the cell may be modulated by a barrier that is disrupted by acid hydrolase activity.

FACTORS IN THE PROGRESSION OF FIBRO-GELATINOUS LESIONS

In human material the sequence of events can only be guessed at, but there seems to be general agreement that the low-lipid, gelatinous elevation represents an early stage, that this may be followed by disruption of the linear collagen fibres in the deep layers of the lesion to give a network or honeycomb of fibres that become associated with fine perifibrous lipid droplets, the peri-fibrous lipid may increase to give a dense coating and SMCs disappear from the central area, and eventually the central pool of atheroma lipid develops in some, but not all plaques. (12,14,7).   In some lesions no fat-filled cells can be observed, but others may contain considerable numbers of fat-filled round cells, most frequently located at the corners of the atheroma lipid pool, but sometimes extending over the top of it as well;   these are probably mainly mononuclear phagocytes or macrophages (19).   These morphological changes are paralleled by distinctive biochemical changes.

Changes in Lipids and LDL

The lipids that accumulate in lipid-rich gelatinous lesions are very similar to the lipids of plasma LDL, but with increasing lipid accumulation there is a decrease in the proportion of phospholipid.   The major component is cholesterol ester predominantly esterified with linoleic acid (18:2), which again suggests that the lipid is derived directly from plasma LDL (3).

Gelatinous lesions contain large amounts of LDL and fibrinogen (Table 1), confirming that they are indeed insudation lesions.   Compared with normal intima, in the early low-lipid lesions LDL is increased 3-4 fold, and 88% is freely mobile in an electric field. However, as the lipid content of the lesion increases the free LDL fraction tends to decrease while there is a highly significant increase in a tightly bound lipoprotein fraction that can be released by incubation with proteolytic enzymes (20).

Table 1.   Free and Bound LDL, Fibrinogen and Fibrin in Normal
Intima and Developing Gelatinous Lesions

	CONCENTRATION: mg/100 mg dry tissue				
	Free LDL	Bound LDL	Fibrin-ogen	Fibrin	Residual Cholesterol [Ø]
Normal intima (n = 27)	3.8	0.6	2.5	2.5	3.6
Gelatinous lesions: Low lipid: whole depth (n = 25)	13.7	1.8	7.7	4.3	5.4
Lipid rich: centre only (n = 35)	5.4	7.4	7.7	22.7	89.6

[Ø]   Cholesterol remaining in the tissue after removal of lipoproteins
by electrophoresis.

       In addition to becoming tightly bound the LDL molecule appears
to undergo physical change in the developing plaque. Smith et al. (21)
combined first dimension molecular sieving and isoelectric focusing
in 3.3% polyacrylamide gels and second dimension immunoelectrophoresis
into antisera to apo-B and apo-C.   In tissue samples there was a component that focused with plasma LDL (Table 2, peak 2) but also a component of large molecular size that failed to enter the 3.3% acryl-
amide gel (Table 2, peak 1).   In normal intima and low lipid gelatinous lesions the major component behaved like plasma LDL, but with increasing accumulation of cholesterol there was a decrease in LDL and
increase in the large, peak 1 component.   In lipid-rich lesions there
was no LDL and all the lipoprotein - both free and tightly bound - was
in the form of the peak 1 component (Table 2).   From its behaviour
with antisera to apo-B and apo-C the peak 1 component seems to be an
aggregated form of LDL and not either VLDL or IDL (21).

       Thus within the developing gelatinous lesion there is a high
concentration of LDL that, in association with accumulation of cholesterol, undergoes both aggregation and tight binding.   It must also
be destroyed, with deposition of the insoluble lipid moiety.   Although
LDL is remarkably resistant to proteolytic attack (22,23) it is
destroyed quite rapidly in incubated intima at low pH by an enzyme
with the characteristics of a lysosomal cathepsin (24).   It is possible that limited cptheptic digestion induces aggregation (21) but
this requires further investigation;   all the tightly bound lipoprotein seems to be in the aggregated form (21) but it is not clear if
aggregation necessarily precedes binding.   In the lipid-rich centres

Table 2.   Apo-B-containing Lipoproteins Separated by Isoelectric
Focusing: Comparison of Normal Intima and Low and High
Lipid Gelatinous lesions

| | LDL components: % distribution | | | Residual |
	Peak 1: origin	Peak I*	Peak 2: LDL	Cholesterol: mg/100 mg.
Normal intima	26.3	5.1	68.6	3.5
GELATINOUS LESIONS				
Low lipid ∅	25.5	12.3	62.2	6.0
Abundant PFL	40.8	11.4	48.9	15.3
Massive PFL	100	NIL	NIL	84.6

∅ PFL = perifibrous lipid   * Intermediate peak.

of lesions bound lipoprotein was destroyed seven times faster than
the free fraction, thus binding of LDL may be of major importance in
the accumulation of lipoprotein-derived extracellular lipid (24).

Fibrin in Lesions and its Relation to Binding of LDL

It can be seen in Table 1 that, compared with normal intima,
there is a three-fold increase in fibrinogen in the low-lipid gelatinous lesions whereas there is a much smaller (and not significant)
increase in insoluble fibrin. By contrast, in the lipid-rich areas
fibrin is increased tenfold, and the concentration of bound LDL is
increased more than ten-fold.

In our studies on release of the bound lipoprotein fraction (20)
we consistently found that the most effective purified enzyme was
plasmin, which is the physiological fibrinolytic enzyme. Equal
amounts were released by a crude collagenase preparation with general
protease activity, but not by purified collagenase, and this, together with the high levels of fibrin in areas of gelatinous plaques that
were rich in both residual cholesterol and bound lipoprotein, suggested that lipoprotein might be bound to fibrin. If it is the bound
lipoprotein that is most rapidly degraded with deposition of its
lipid moiety, then this implies that fibrin plays a key role in
lesion development.

However, the relation between fibrin and bound lipoprotein is
not a simple one; it can be seen in Table 3 that non-endothelialized
thrombi do not bind LDL (25). There was substantial binding in
fibrin mural thrombi that were covered with endothelium and showed
some invasion of collagen from the base, although these lesions had
accumulated only small amounts of residual cholesterol. The

Table 3.   Comparison of Free and Bound LDL and Fibrinogen in
Gelatinous and Fibrous Plaques, Thrombi and Prosthesis
Graft Pseudo-intima

	Concentration: mg/100mg dry tissue				
	Free LDL	Bound LDL	Fibrin- ogen	Fibrin	Residual Cholesterol
Lipid-rich areas of plaques:					
Gelatinous	5.4	7.4	7.7	22.7	89.6
White fibrous	1.5	3.3	1.9	4.9	72.9
Thrombi:					
Free-lying fibrin straps: no endoth- elium	1.3	0.3	0.6	83.8	1.5
Mural thrombi: no endothelium	2.0	NIL	2.7	74.3	2.6
endothelium + collagen invasion	10.1	4.3	21.9	47.7	5.3
"Encrustations" on ulcerated plaques	3.8	6.6	6.4	57.4	119.3
Aortic graft pseudo- intima: no endothelium	1.5	0.2	1.8	87.2	2.4

familiar "cauliflower-like encrustations" on ulcerated plaques also
contained large amounts of bound lipoprotein; these structures contain masses of flat cholesterol crystals, packed on edge, and covered
with endothelium and one or two thin strands of collagen.   We do not
really know if the LDL binding is primarily associated with invasion
of collagen or with the presence of endothelium.   In the non-endothelialized free lying fibrin thrombi, mural thrombi and prosthesis
graft pseudo-intimas (in which there was no invasion of collagen or
SMCs) the concentrations of bound LDL were very low, but so were the
concentrations of free LDL and fibrinogen.   In some way that we do
not understand endothelialization of thrombi seems to promote accumulation of LDL and fibrinogen.

Origin of the Fibrin and its Role in Plaque Development

The Velicans did not see evidence of mural thrombus initiating
lesion development in young age groups (7) and this is supported by

Table 4.   Rate of "clotting" of Fibrinogen and Concentration of
            Clotting Factors in Normal Intima and Lesions

| | FIBRINOGEN CLOTTED | | Prothrombin | MOLAR RATIO |
	mg/100mg∅ in 3 hours	% of total	concentration μg/100mg∅	Prothrombin/ Antithrombin III
Normal Intima	0.3	20	43.4	0.3
Gelatinous	1.5	25	72.1	0.4
Plaque centres	2.0	76	75.6	1.1
Mural thrombi	5.9	70	110.8	0.6

∅ mg lipid-extracted dry tissue

the chemical finding of low concentrations of fibrin in most low-lipid
gelatinous thickenings (Table 1).   Above age 30 they found a small
proportion of microthrombi associated with developing plaques, while
Haust reported strands of fibrin in the insudate in larger lesions
(12).   Both mural thrombosis and formation of fibrin from fibrinogen
within the intima are compatible with our biochemical and morphological findings.   In a comparison of samples of intima minced and incubated with saline there was loss of fibrinogen compared with samples
treated with EDTA, and this apparent "clotting" was most active in the
centres of plaques (Table 4).   Using the immunoelectrophoretic assay
we also found specific retention of prothrombin in plaque centres, and
a fourfold increase in the molar ratio of prothrombin to antithrombin
III, which is probably the principle inhibitor of thrombin (Table 4).
Thus the components for clotting of fibrinogen are potentially present
in the intima although this type of study does not tell us if the
system is activated in non-disrupted tissue in vivo.   Microscopically
we have found thin fibrin mural thrombi that were not recognized
macroscopically;   they occur most frequently on the gelatinous "tails"
of plaques, but also occasionally on isolated gelatinous thickenings.
There is no evidence that these fibrin deposits are associated with
pre-existing fissures in underlying plaques.

"Organization" of arterial thrombi takes many different forms,
ranging from invasion by masses of leucocytes, to fibrin networks
containing virtually no leucocytes, in which migration of SMCs and
deposition of collagen along the fibrin strands can be clearly seen.
They produce an ordered, linear structure that is indistinguishable
from the cap of a fibrous plaque.   These thrombi are also associated
with the tight-binding of LDL (Table 3), suggesting that binding may
occur at the fibrin-collagen interface.   A fibrin network seems to
stimulate migration - and probably proliferation - of SMCs;   if it
also leads to the tight binding of LDL it may play a key role in the
genesis of plaques.   The role of haemostatic factors in athero-

genesis is an area that has been greatly neglected in recent years although in 1973 Sumiyoshi et al. (26) demonstrated the development of typical fibrous atherosclerotic plaques with fibrous caps and lipid-rich centres from non-occlusive mural thrombi in the aortas of normolipidaemic, normocholesterolaemic rabbits. Recent prospective epidemiological studies have shown that fibrinogen and some clotting factors are significant predictors of ischaemic heart disease (27).

CALCIFICATION AND ULCERATION OF PLAQUES

My brief was to discuss the development of lesions from the fatty streak to calcified plaque, but in this last stage I have to admit failure. Meyer and Lind (28) showed patchy calcification of the IEL in iliac arteries of newborn babies and infants, and there is a progressive increase in the calcium content of arterial elastin with age (29). In old aortas and lesions it seems to be associated with the binding to elastin of highly cross-linked collagen (29). However, elastin is not a prominent component of most human lesions, and despite elegant studies on the composition, crystalline form and microscopic architecture of calcium deposits (30,31) there seems to be very little information on the factors that lead to deposition of massive plates of hard calcium in atherosclerotic lesions. Calcium deposition is not necessarily linked to lipid deposition (30), but it is not clear to what extent lipids are a factor in the specific deposition of calcium in lesions.

In terms of clinical sequelae the deposition of plates of calcium are clearly of major importance, drastically changing the rheology of the vessel wall, while the edge of a calcium plate is a frequent site of ulceration. I have no doubt that ulceration is a major cause of thrombotic occlusion in both coronary and peripheral arteries. The extent of ulceration shows a bewildering range - aortas with massive proliferative lesions in which none is ulcerated, others with comparable lesions in which almost every one is ulcerated; aortas in which the intima is almost completely disintegrated in the lower abdominal segment while there are no ulcers in thoracic segment, and others with one significant plaque only, located in the upper abdominal or thoracic segments but deeply ulcerated. The causes of this variation pose a fascinating problem which, to the best of my knowledge, has never been investigated; perhaps I can encourage a member of this Study Institute to take up the challenge.

CONCLUDING REMARKS

In this talk I have concentrated mainly on the factors that may be involved in the progression of the early, low-lipid gelatinous lesion into the typical fibrous plaque with lipid-rich centre that is generally accepted as the significant lesion in occlusive vascular disease, and have tried to emphasize the key role that may be played by fibrin.

However, I believe that the <u>central</u> event is proliferation of
SMCs.   Insudation may regress if there is no SMC proliferation, but
once there has been a proliferative response reduction of the always
precarious oxygen supply will lead to accumulation of lactic acid and
leakage and activation of lysosomal acid hydrolases.   The idea of
localized release of a growth factor from platelet microthrombi is
attractive, but Dr. Ross emphasized that there is no evidence that
this happens <u>in vivo</u>, and the Velicans' study does not support either
this idea or the idea of growth into fibrin microthrombus.   We still
have much work to do before we understand the stimulation of localized SMC proliferation in the large arteries of a normal, intact man.

REFERENCES

1.   R.L. Holman, Atherosclerosis - a pediatric nutrition problem?
      <u>Amer.J.Clin.Nutr.</u>, 9: 565 (1961).
2.   H.C. McGill, Fatty streaks in the coronary arteries and aorta,
      <u>Lab.Invest.</u>, 18: 560 (1968).
3.   E.B. Smith, The relationship between plasma and tissue lipids in
      human atherosclerosis, <u>Advan.Lipid Res.</u>, 12: 1 (1974).
4.   D. Velican and C. Velican, Study of fibrous plaques occurring in
      the coronary arteries of children, <u>Atherosclerosis</u>, 33: 201
      (1979).
5.   D. Velican and C. Velican, Atherosclerotic involvement of the
      coronary arteries of adolescents and young adults, <u>Athero-
      sclerosis</u>, 36: 449 (1980).
6.   C. Velican and D. Velican, Incidence, topography and light-
      microscopic features of coronary atherosclerotic plaques in
      adults 26-35 years old, <u>Atherosclerosis</u>, 35: 111 (1980).
7.   C. Velican and D. Velican, The precursors of coronary athero-
      sclerotic plaques in subjects up to 40 years old, <u>Athero-
      sclerosis</u>, 37: 33 (1980).
8.   E.B. Smith, Metabolism of the arterial wall, this volume p
9.   J.L. Goldstein and M.S. Brown, The low-density lipoprotein
      pathway and its relation to atherosclerosis, <u>Ann.Rev.Biochem.</u>,
      46: 897 (1977).
10.   J.L. Goldstein, R.G.W. Anderson, L.M. Buja, S.K. Basu and M.S.
      Brown, Overloading human aortic smooth muscle cells with low
      density lipoprotein-cholesteryl esters reproduces features
      of atherosclerosis in vitro, <u>J.Clin.Invest.</u>, 59: 1196 (1977).
11.   R.W. Mahley, Cholesterol feeding: effects on lipoprotein
      structure and metabolism, <u>in</u> "Atherosclerosis V"., A.M. Gotto,
      L.C. Smith and B. Allen, eds., Springer-Verlag, Berlin, 641
      (1980).
12.   M.D. Haust, The morphogenesis and fate of potential and early
      atherosclerotic lesions in man, Human Pathol., 2: 1 (1971).
13.   E.B. Smith and R.S. Slater, Lipids and low density lipoproteins
      in intima in relation to its morphological characteristics,
      <u>in</u>, "Atherogenesis - Initiating Factors", Ciba Foundation Sym-
      posium No. 12, New Series, Excerpta Medica, Amsterdam, 39 (1973).

14.   E.B. Smith and R.H. Smith, Early changes in aortic intima,
      Atheroscler.Rev., 1: 119 (1976).
15.   E.B. Smith, Acid glycosaminoglycan, collagen and elastin content
      of normal artery, fatty streaks and plaques, in, "Arterial
      Mesenchyme and Arteriosclerosis", W.D. Wagner and T.B. Clark-
      son, eds., Advan.Exp.Med.Biol., 43: 125 (1974).
16.   E.B. Smith, Relation between arterial wall and risk factors for
      coronary heart disease: lipoproteins - steady state aspects,
      in, "Atherosclerosis IV", G. Schettler, Y. Goto, Y. Hata and
      G. Klose, eds., Springer-Verlag, Berlin, 24 (1977).
17.   R. Ross, A. Vogel, E. Raines and B. Kariya, The platelet-derived
      growth factor, in, "Atherosclerosis V", A.M. Gotto, L.C.
      Smith and B. Allen, eds., Springer-Verlag Berlin, 442 (1980).
18.   E.P. Benditt, Implication of the monoclonal character of human
      atherosclerotic plaques, Amer.J.Pathol., 86: 693 (1977).
19.   C.W.M. Adams and O.B. Bayliss, Detection of macrophages in
      atherosclerotic lesions with cytochrome oxidase, Brit.J.Exp.
      Pathol., 57: 30 (1975).
20.   E.B. Smith, I.B. Massie and K.M. Alexander, The release of an
      immobilized lipoprotein fraction from atherosclerotic les-
      ions by incubation with plasmin, Atherosclerosis, 25: 71
      (1976).
21.   E.B. Smith, H.S. Dietz and I.B. Craig, Characterization of free
      and tightly bound lipoprotein in intima by thin-layer iso-
      electric focusing, Atherosclerosis, 33: 329 (1979).
22.   P. Bernfeld and T.F. Kelley, Proteolysis of human serum β-
      lipoprotein, J.Biol.Chem., 239: 3341 (1964).
23.   M.J. Chapman, S. Goldstein and G.L. Mills, Limited tryptic
      digestion of human serum low density lipoprotein: isolation
      and characterization of the protein-deficient particle and
      of its apoprotein, Eur.J.Biochem., 87: 475 (1978).
24.   E.B. Smith and I.B. Massie, Destruction of endogenous low
      density lipoprotein in incubated intima, Atherosclerosis,
      26: 427 (1977).
25.   E.B. Smith, E.M. Staples, H.S. Dietz and R.H. Smith, Role of
      endothelium in sequestration of lipoprotein and fibrinogen
      in aortic lesions, thrombi and graft pseudo-intimas, Lancet,
      ii: 812 (1979).
26.   A. Sumiyoshi, R.H. More and B.I. Weigensberg, Aortic fibrofatty
      type atherosclerosis from thrombus in normolipidemic rabbits,
      Atherosclerosis, 18: 43 (1973).
27.   T.W. Meade, R. Chakrabarti, A.P. Haines, W.R.S. North, Y.
      Stirling and S.G. Thompson, Haemostatic function and cardio-
      vascular death: early results of a prospective study,
      Lancet, i: 1050 (1980).
28.   W.W. Meyer and J. Lind, Calcifications of the carotid siphon -
      a common finding in infancy and childhood, Arch.Dis.Child-
      hood, 47: 364 (1972).

29. S.M. Partridge and F.W. Keeley, Age related and atherosclerotic changes in aortic elastin, in, "Arterial Mesenchyme and Arteriosclerosis", W.D. Wagner and T.B. Clarkson, eds., Advan.Exp.Med.Biol., 43: 173 (1974).

30. S.Y. Yu, Calcification processes in atherosclerosis, in, "Arterial Mesenchyme and Arteriosclerosis". W.D. Wagner and T.B. Clarkson, eds., Advan.Exp.Med.Biol., 43: 403 (1974).

31. K. Schmid, W.O. McSharry, C.H. Pameijer and J.P. Binette, Chemical and physicochemical studies on the mineral deposits of the human atherosclerotic aorta, Atherosclerosis, 37: 199 (1980).

PROGRESSION AND REGRESSION OF ADVANCED ATHEROSCLEROSIS AS STUDIED

BY QUANTITATIVE METHODS

Robert W. Wissler

The Department of Pathology and
The Specialized Center of Research in Atherosclerosis
The University of Chicago, Chicago, Il. U.S.A.

INTRODUCTION

In this presentation, the evidence will be reviewed which indicates that the incidence and severity of atherosclerosis and its clinical effects can be altered significantly by dietary and other factors. The main goals will be to put this evidence in perspective, to call attention to its shortcomings and then to outline the studies that appear to be most urgently needed. Before considering the ways one may measure progression and regression of atherosclerotic plaques, it is important to understand the main features of atherosclerosis in humans.

The typical advanced atherosclerotic plaque is made up of two principal components. One of these is the necrotic core which is filled with a grumous mixture of cholesterol, cholesterol esters, neutral lipids and proteins. The atheroma gets its name from this soft center. The second major component is the fibrous cap which is made up mostly of smooth muscle cells and their products, i.e., collagen, elastin and proteoglycans, but which consistently contains variable and often substantial amounts of intracellular and extracellular lipid[1,2,3]. Chemically, in those human lesions that have been analyzed, lipids, especially cholesterol, cholesterol esters and triglycerides, are the predominant components along with the fiber proteins[4].

The other components such as calcium, fibrin and inflammatory cells appear to be largely secondary to the developing plaque. Generally speaking, they do not contribute substantially to its size or progression. In fact, when calcium is a prominent

component it is usually because it displaces much of the necrotic
area. On the other hand, some of the minor elements may have an
influence far beyond their quantity. For example, proteoglycans
may be particularly important in binding low density lipoproteins
in the artery wall[4] and the macrophage population, although small,
may furnish powerful peptides and enzymes which may, in turn,
contribute to cell proliferation, collagen dissolution, lipoprotein processing, etc.[5].

Recent investigations have rather firmly established that
the smooth muscle cells of the lesions are probably derived from
the arterial media by a combined process of migration and proliferation[6,7]. The collagen, elastin, and proteoglycans are almost certainly synthesized by these cells[8,9]. The cholesterol,
cholesterol esters, and proteins present interstitially are
partly derived from intracellular lipid and protein liberated
by the necrosis of preexisting cells and partly from the accumulation of blood proteins and lipoproteins that make their way
into this lesion from the lumen of the artery[2].

The proliferated smooth muscle cells and their products,
along with the lipid-rich necrotic area, form the principal space-
occupying components of the lesions and are responsible for the
stenosis of the coronary, carotid, and femoral arteries that frequently lead to ischemic complications such as myocardial infarction, and ischemic gangrene of the extremities. Rupture or fracture of the fibrous cap and the thromboplastic character of the
underlying components including the lipids in the necrotic core
of the lesion also trigger the process of arterial thrombosis,
which can lead to sudden occlusion of these arteries.

As the disease becomes more severe, it may involve more and
more of the media of the artery until the artery wall, especially
the aorta, is weakened and undergoes a remarkable saccular dilation in the area of the advanced soft, grumous plaque, thus
leading to an aneurysm that may be several times the diameter
of the original artery.

Atherosclerosis should be differentiated from other harden-
ings and thickenings of the artery wall that do not contain abundant lipid (cholesterol and cholesterol esters) and that do not
lead to remarkable narrowing of the lumen of the medium-sized
arteries. One of the responsibilities of the medical and scientific community is to distinguish clearly between many types
of spontaneous and experimental lesions of the artery wall of
animals that have little or no resemblance to the atherosclerotic
process and those which serve as excellent models of the human
disease. One of the aims of this brief review is to help the interested reader make that differentiation.

PATHOGENESIS OF PLAQUES

During the past two decades there has been a remarkable increase in knowledge concerning the pathogenesis of atherosclerosis. Much of this developing understanding of the atherosclerotic
process has been summarized in recent reviews[10,11,12]. It also appears that arterial medical smooth muscle cell migration from
media to intima and a definite increase in the intimal proliferation of these cells are among the earliest parts of the process.
It is becoming evident that endothelial damage may help facilitate
the migration of macromolecules of low-density lipoproteins into
the artery wall and may encourage the phenomenon of platelet
sticking, which may in turn furnish one of the stimuli for arterial smooth muscle cell proliferation[11,12]. On the other hand,
it is now evident that one episode of even severe endothelial
damage will not sustain a progressive proliferation of arterial
wall cells unless the blood lipid levels are abnormally high[13] and
that a considerable degree of sustained platelet adhesion to denuded areas of the artery wall can be present without progressive
arterial smooth muscle cell proliferation and plaque formation[14].
What is not known is whether a plaque usually forms without more
than the usual "wear and tear" hemodynamic endothelial damage and
whether the average individual has a large or small component of
endothelial injury as part of the underlying pathogenesis.

Equally puzzling is the question of whether there is more
than one kind of early lesion[1], and if so whether one or more of
these is more likely to be progressive than others[15]. Smith and
Slater[16] and Panganamala et al.[17] have proposed that the classical
type of fatty streak in which most of the lipid is intracellular
is not the precursor of the usual progressive atheromatous
plaque. A part of this evidence is based on the differing topography of the fatty streak and the progressive plaques in the
aorta[18], and some of it is based on the rather remarkable differences in the biochemical analyses of the cholesterol esters
of the fatty streak and the progressive plaques as well as the
differing lipoprotein/apoprotein contents of these two lesions[16].
Since fatty streaks are present in most human populations and
in most species of experimental animals whereas progressive
atherosclerotic plaques are found in only a few human populations and only under special circumstances in experimental animals, there appears to be a real need for further definition
of the early determinant of the progressive atherosclerotic
plaque. Recently preliminary observations that we have reported
indicate a rather remarkable variability in the fatty streaks
found in human subjects who die suddenly at the end of the
second decade of life. In some, most of the lipid is intracellular; in others, it is mostly extracellular; and in still
others, there is a rather equal mixture of intracellular and

extracellular lipid in small flat lesions that are indistinguishable grossly. Some flat lesions have much more involvement of
the arterial media than others. Futhermore, some lesions occur
in which most of the lipid is bound to elastic fibers or to
other extracellular components of the lesion. These observations indicate that further study of the "fatty streak" is
necessary to define those components that are most likely to
accompany progression[15].

Another enigma of the pathogenesis of atherosclerosis is the
rather remarkable variability of the components of cell proliferation and necrosis, as well as fiber protein and glycosaminoglycans
deposition, along with the variability in the prominence of lipid
deposition in the advanced plaque. Experimental results in non-
human primates may ultimately give some insight into the remarkably varied histopathology that is observed in the advanced human
plaque[19,20,21,22].

EXPERIMENTAL MODELS AND THE NATURE OF EVIDENCE THAT IS DERIVED
FROM THE STUDY OF ATHEROSCLEROSIS PROGRESSION AND REGRESSION

Most of the early work on the pathogenesis of atherosclerosis involving animal models was descriptive[23,24,25]. The
lesions were generally produced by feeding diets designed to raise
the serum cholesterol levels and most of the emphasis was placed
on evaluating the presence of stainable lipid, either grossly or
microscopically. Most of the evaluations were descriptive. When
quantitation was attempted it usually took one of two forms.
Either the investigator used an estimation or a ranking of the %
of surface area involved by grossly discernable lesions based on
visual inspection of the aorta by two or more observers or by
planimetry, or he/she used an estimation of frequency and/or
severity of lesions based on microscopic study of the lesions
sampled in some standardized way and usually evaluated on a
1-4+ scale.

As more has been learned about the comparative pathology of
atherosclerotic lesions in various species and under varying
dietary conditions[26,27] it has become evident that the components and the topography vary greatly from model to model. There
has been a remarkable increase in sophistication of study of
lesion components and quantitation of lesion severity which will
become evident later in this presentation.

THE NATURE OF THE EVIDENCE THAT HAS BEEN USED TO DEVELOP THE
CONCEPT OF HUMAN PLAQUE REVERSIBILITY

Various lines of evidence developed over the past 60-70
years, from human studies, indicate that atherosclerotic lesions

and their effects can be favorably influenced. The first line of
evidence is primarily based on changes in attack rate and/or mortality, i.e. documenting, usually retrospectively and longitudinally,
a reduction in the clinical effects of atherosclerosis when the
population of a nation has its food supply drastically decreased,
as has often happened during war time. This type of evidence is
also obtained when a large group of subjects is studied prospec-
tively and/or half of them are treated, for example with a steroid
hormone or with endocrine gland ablation that alters serum lipids
and lipoproteins greatly.

There are problems in evaluating the significance of the
results of these types of studies. For example, when mortality
is the endpoint and autopsy results are not available then one
cannot be certain whether an observed effect is due to an effect
on the atheromatous lesions or simply due to a decrease in the
clinical effects of the lesions - effects such as thrombosis or
severity of heart attack, etc. Furthermore, the studies are often
poorly controlled, especially when they are "longitudinal" and/or
"retrospective".

In spite of these problems it is valuable to look at results
such as those developed by Malmros and summarized by Schettler
from the observations in Sweden during World War II[28]. Here the
reduction in dietary fat was accompanied by a very sharp reduction
in cardiovascular deaths. Many scholars believe that this was
due to a decrease in thrombosis - not due to regression of atherosclerosis because it appears to be too sudden to allow time for
regression. But this was not an isolated phenomenon. Data from
three countries, Sweden, Finland and Norway, has recently been presented by Malinow[29], which indicate that the atherosclerosis
mortality in populations of all these countries was similarly affected and in general the trends were well timed to reflect the
changes in dietary fat. At the same time there was a rising incidence of heart attack deaths in the USA where no such changes
in food supply occurred.

Can these kinds of results be obtained in studies where
the serum cholesterol is purposely lowered? When a bile acid
sequestering agent (colestipol) was administered recently to half
of the subjects in a clinical trial, it produced very substantial
reductions in serum cholesterol values. The results show striking
decreases in both atherosclerosis mortality and general mortality[30].
Although attempts have been made to discount this study[31], it is
probably one of the least flawed of the modern relatively large-
scale cholesterol-lowering clinical trials that have been reported.

Similarly, rather remarkable results have been reported recently by Turpeinen from Finland[32] and by Leren in Norway[33] where
nutritional measures are being used to try to lower the attack

rate from atherosclerosis. Although neither the serum lipid -
lowering effects nor the changes in mortality are quite as
striking as in the colestipol studies, there is no doubt that they
indicate that even with moderate dietary alterations and serum
lipid reductions mortality can be beneficially affected.

REDUCTION IN SEVERITY OF THE DISEASE PROCESS

The second general approach has been to study the reductions
in frequency or severity of atherosclerosis at autopsy - again
often retrospectively and longitudinally, when nutritional elements are reduced substantially, usually during war time or in
other studies when the autopsy evaluation of atherosclerosis in
subjects with wasting diseases such as cancer or chronic infec-
tion is compared with that in patients who die with non-wasting
disorders[29,34]. More recently the severity of atherosclerosis
has been compared in subjects who have been treated with drugs or
cholesterol lowering diets or surgery such as "partial ileal
bypass"[35,36].

In the past when autopsy results have been used to make these
comparisons retrospectively, serious flaws developed in the studies.
The methods used to grade or evaluate changes in atherosclerosis
were often non-quantitative and subjective. Furthermore,
microscopic study of standardized areas of arteries was not utilized. This approach is valuable because of its superior relia-
bility and its greater power of documenting what has happened to
the components of the lesions.

In spite of the shortcomings of these older studies, it is
interesting that Finland also offers some of the best evidence
that atherosclerosis was reduced when lesions were studied in
longitudinally compared groups (1933-38 and 1940-46)[37]. It is
notable that the more advanced lesions appear to be the most
affected and that the younger and middle-aged persons who died
were more likely to show more definitive differences.

Somewhat similar trends that may be related to nutritional
changes were noted in sudden accidental death autopsies of Swiss
Soldiers[32] - although the reductions where not necessarily due to
lesion regression. The results of the Marburg Study reported by
Schettler[38] are much more likely due to regression of lesions.
Nevertheless all of these studies suffer from their longitudinal,
retrospective designs and the lack of precision in their methodology.

The great and perceptive New York pathologist, Sigmund Wilens,
reported, a little over 30 years ago[34], probably the most careful
study of autopsy generated data, again retrospective, but with
rather striking results. There was a remarkable correlation

between nutritional state and the incidence of advanced atherosclerosis, especially in the upper age groups 65 to 75+ years,
and a concomitant rise in the cases with little or no evident
atherosclerosis in the average to poor state of nurtition autopsies in the same age group. He also reported that the lipid rich
lesions seemed most susceptible to regression, a finding which
is similar to our results in the monkey.

There are a number of studies, and the results of only a few
can be mentioned here indicating by means of autopsy data that
therapeutic ventures on one kind or another can be associated
with changes in the severity of atherosclerosis. There are two
such well documented studies, dealing with the effects of estrogens
or of ovariectomy, one from New York and one from Los Angeles.
The one reported by Rivin et al.[39] indicates as experimental results have in our laboratory[40] and in others[41], that oophorectomy
in women is likely to result in more severe atherosclerosis while
the administration of estrogen generally is associated with decreased atherosclerosis. In addition, this group found that
males with carcinoma of the prostate treated with estrogen showed
far lower incidence of atherosclerosis in the coronary and cerebral arteries and in the aorta than did the controls who received
little or no estrogen[42]. Similar autopsy results, have been
described in males by London et al.[42] mostly in those subjects
treated with estrogen for carcinoma of the prostate.

The results that Henry Buchwald and his colleagues have
been reporting from Minneapolis are even more promising[43]. These
veteran investigators have developed an especially effective and
apparently safe surgical procedure to lower the serum cholesterol
level by diverting the majority of bile acids from being reabsorbed. This operation, called "partial ileal bypass" is
not only as effective as the "bile acid sequestering agents,"
cholestyramine and colestipol, but it is in a sense more de-
pendable since patient adherence is not a factor. Dr. Buchwald
has recently reported[35] that as compared to other groups of
patients controlled only by diet and studied at other centers, a
group of high risk patients subjected to "partial ileal bypass"
and studied by sequential coronary arteriography had many fewer
lesions showing progression and that surgery resulted in a very
high percentage of patients whose coronary lesions either remained stationary or definitely regressed in a period of about
18 months. This study is continuing as a consortium study involving several Medical Centers and with adequate controls.

All of these studies provide definite evidence that chronic
weight loss and low serum lipid (cholesterol) levels resulting
from low fat and low calorie nutrition or from wasting disease
produce regression of atherosclerosis and/or reduce its clinical consequences, including fatal heart attacks; they provide

substantial evidence (from several studies) supporting the view that lowering serum lipids (cholesterol) by diet, drug or surgery results in regression or at least marked retardation of the progression process in the majority of cases.

It is also evident that these studies leave us with a number of problems that need to be solved in order to provide more meaningful data. These problems can be simply stated to be:

1. Lack of definitive quantitative measurements
2. Lack of adequate controls
3. Lack of prospective studies that include autopsy data
4. Lack of microscopic studies of regression of atherosclerosis in humans

Some of these kinds of problems can be much more adequately dealt with in experimental studies. In fact, if one has confidence in the experimental model of atherosclerosis, then well-controlled, quantitatively evaluated prospective studies that include morphometric microscopic analysis or biochemical analyses and the additional information that they provide can be most illuminating.

PROGRESS AND PROBLEMS WITH ANIMAL MODEL STUDIES OF REGRESSION OF ATHEROSCLEROSIS

The early studies of treatment of atherosclerosis were carried out in rabbits and many of them were said to provide evidence of increasing severity of lesions rather than improvement[44]. This paradoxical response may be related to the large stores of cholesterol that the rabbit deposits in its reticuloendothelial system when it is fed a high cholesterol ration. These over-filled pools are depleted very slowly during regression. These features and the tendency for its atheromatous lesions to contain rather large proportions of lipid-laden macrophages may contribute to the resistance it has shown to regression of its advanced plaques[45,46]. These failures of the rabbit model probably delayed advancement in this important field for many years[47,48]. Now however, it is generally realized that the rabbit model, as it is usually employed, is dissimilar to atherosclerosis in human subjects in several important respects[49,50,51] and a few studies have been recorded which appear to demonstrate that with rather complex approaches the process will regress remarkably[52,53,54].

The rhesus monkey studies by Taylor et al.[55] opened a new era for the study of atherosclerotic disease, including regression. Essential gross and microscopic features of advanced human atherosclerosis are produced in the rhesus model. It also shows some of the usual "clinical" effects of the lesions, such as myocardial infarction[56] and ischemic damage in the extremities[57].

Armstrong, Warner and Connor[58] first reported substantial regression of rather advanced atherosclerosis in the rhesus model. Male monkeys, after developing stenosing coronary artery lesions, were treated with a diet which was either low fat, essentially cholesterol-free, or high in polyunsaturated fat (corn oil) and very low in cholesterol. A substantial decrease in the stenosis was observed in a 40 month treatment period along with impressive changes in the plaques. They became flatter, firmer and smoother and thus probably less likely to produce ischemic damage or to stimulate thrombosis[59]. Armstrong and Megan [60] also demonstrated substantial decreases in cholesterol (especially cholesterol esters) and other lipids in the plaques. As therapy was continued on an anatomical area or unit basis, collagen was also found to be decreased substantially[61].

Our studies of the process of regression of lesions in the male rhesus monkey have shown that a definite decrease in lesion size, largely due to a reduction in cholesterol ester, can occur in 12 to 18 months[47,59,62]. More recently substantial plaque regression has also been demonstrated at 8 months[63]. Intracellular and extracellular lipid components along with the necrotic center are remarkably reduced when the serum cholesterol concentrations have returned to baseline levels for 4-6 months. Accompanying these changes in visible lipid, substantial condensation and remodeling of collagen occurs; the endothelium, which is badly damaged in advanced disease, undergoes almost complete "healing"[64,65]. The amounts of microscopically stainable lipid which remain after therapy seem to be almost entirely localized in areas rich in glycosaminoglycans and/or elastin[47,63].

Even if a very atherogenic diet is fed during therapy and serum cholesterol levels remain somewhat elevated, lesions in the rhesus monkey can be successfully treated with cholestyramine, a bile acid sequestering agent[47,59]. The three consecutive experiments that we have completed recently indicate that coronary lesions are somewhat more resistant to therapy than are those in the aorta. Diet and cholestyramine seem to act synergistically to affect the cholesterol turnover rate[47,66]. Point counting morphometry extended to include computer assisted digitizer morphometric studies, show that the necrotic areas as well as the extracellular and intracellular lipids disappear[67,68,69].

The main variables among these three consecutive experiments that have been performed using this plan have been the type of food fats fed in order to produce the advanced atherosclerotic disease and the type of diet which has served as a therapeutic diet during the second half of the study, but the results have been very similar in each experiment.

The results of studies of this kind can be quantitated and

expressed in many ways. The gross involvement of the intima
by raised or by lipid-containing lesions in unstained specimens
or in specimens stained for fat can be evaluated reproducibly
using a number of planimetric or point counting approaches[70],[71].
In our most recent evaluation of the microscopic samples of the
lesions in our studies, we have utilized point counting[68], the
thickness of intimal lesions based on an actual micrometer
measurement in standard sections of the thoracic and abdominal
aorta and a number of parameters of lesion size such as lumen
area, necrotic center size, fibrous cap size, the area taken up
by intracellular or extracellular lipid, etc. It is evident
that the two groups being treated with the low-cholesterol, low-
fat ration with or without cholestyramine reveal the most evidence of regression. The group which is fed the atherogenic
ration but also receives cholestyramine shows an intermediate
degree of regression. Although we do not fully understand the
reason for the rather substantial regression when the atherogenic ration is continued and cholestyramine is added to it,
this has been a consistent finding in a number of experiments
including the three that we refer to. Our evidence thus far,
would link this consistent and substantial regression when the
levels of serum cholesterol are substantially above basal concentrations, with a marked increase in cholesterol turnover that
we have found in these animals by means of isotopic studies[47].

A number of additional studies are now on record using this
model and without exception they indicate that the rhesus lesion
has a very strong resemblance to the human plaque and that it
can undergo substantial regression in a relatively short time.
In addition, Professor Giorgio Weber of Siena, working with us,
has demonstrated rather convincingly that along with the plaque
becoming smaller, the arterial endothelium over the plaque
becomes much healthier[65]. In other words, the endothelial
damage found with an advanced plaque heals during regression.
Furthermore, the documentation of this type of favorable reaction
has been confirmed and extended·by Rose Jones in our laboratory
using additional lesion sites and therapeutic approaches[64]. As
regression studies have progressed, Armstrong and Megan have
demonstrated that it may be possible with time to document a real
decrease in fiber proteins, collagen and elastin, in the lesions[61]
and Dr. Radhakrishnamurthy, utilizing standard samples from our
animals has shown some remarkable changes in the various glycosaminoglycans of the lesions following therapy[72].

One swine study deserves mention. Dr. Daoud and co-workers,
at Albany Medical College, utilizing very severe atherosclerotic
plaques, produced in swine by a combination of a high-fat,
high-cholesterol atherogenic ration (with sodium cholate) and
ballon catheter injury have not been able to produce very convinc-
ing evidence of regression in lesions that have large necrotic cen-

ters, but this group has also demonstrated a marked decrease in the rate of plaque cell proliferation when the serum lipids are lowered[73,74].

RECENT HUMAN STUDIES OF REGRESSION OF LESIONS OR DECREASED ATTACK RATE

The colestipol studies in human referred to earlier[30] as well as the cholestyramine studies now underway in the National Heart, Lung and Blood Institute's "lipid clinics"[75] are, in a sense, complementary to these primate and swine studies. So are the partial ileal by-pass trials referred to earlier[43].

There are, regrettably, no human studies to date in which the beneficial effects on the lesions are expressed by truly objective quantitation of autopsy results using modern approaches to morphometry, standardized sampling and pressure perfusion fixation - and if there were, the selective biases in the ways that autopsy permissions are obtained in most industrial-urban societies would make the data difficult to interpret. None of the human studies referred to above have been documented or even illustrated by microscopic study. Thus, they lack the additional insight that histopathological morphometry can give to studies of this kind.

Furthermore, to our knowledge, very few prospective studies have even been planned to include autopsy evidence of regression. This can be done but it is difficult, expensive, and requires extraordinary leadership, planning and cooperation among scientist from several disciplines.

All of these (rhesus monkey) studies have heightened the interest in treating atherosclerosis in people who have advanced disease which has been demonstrated and quantitated by means of arteriography.

The results of this type of study have been reported recently[36] using a computer assisted method of reading arteriograms developed by David Blankenhorn and his co-workers at the University of Southern California[76,77]. These investigators have reported that the evidence of regression of atherosclerosis on sequential femoral arteriography in a group of subjects who were treated with a cholesterol lowering regimen (diet plus medication) was very well correlated with the fall in serum cholesterol observed in the individual.

FUTURE TRENDS

The hope of the future would appear to be the utilization of relatively small, carefully-controlled groups, whose response to therapy will be validated by autopsy-calibrated and autopsy-monitored quantitative, computer assisted, sequential, contrast-media angiography or by non-invasive ultrasound (sonar) angiography combined with Doppler measurements of arterial flow rates[78]. These methods, monitored in representative cases by autopsy interpretation will, we believe, solve many of the problems that we now face with the cumbersome, expensive, large-scale (mass) clinical trials. None of these trials, which really aren't very "large-scale", have tested the effects of intervention on regression of atherosclerotic plaques. The end points have always been clinical events or mortality (presumably from the effects of atherosclerosis) but with no assurance that the fundamental lesion, the atherosclerotic plaque, has been influenced by therapy. Unfortunately most of these studies have been very expensive and many have been flawed in one way or another.

The types of studies that we foresee should be much less complex and expensive and much less likely to have unexpected defects in design. They should be possible in the not too distant future. In fact, both Dr. David Nash, at Syracuse University[79], and Dr. David Blankenhorn, at the University of Southern California[80], have reported surveys of reported cases of patients who revealed reasonably well documented regression of their artherosclerosis on sequential arteriography.

THE MECHANISMS OF REGRESSION

The cell biological and molecular biological mechanisms which are active in regression are largely unknown. It appears probable that when the serum lipids are reduced, including the low density lipoproteins with the heavy load of cholesterol that they carry, then many of the stimuli for cell proliferation and cholesterol ester accumulation will subside[59]. Under these circumstances and given an improved HDL/LDL ratio it appears likely that cholesterol may be removed from the lesions[34]. How much of the apparent decrease in cell replication in the plaque that appears to accompany regression[73] is due to endothelial healing and resultant decrease in platelet derived growth factor activity[81] and in the endothelial cell derived growth factor[82], and how much is due to the lowering of the LDL levels and the decrease in that stimulus for cell proliferation[83] is not known. Neither is it clear, at present, whether some of the mononuclear cells of the plaque elaborate sufficient collagenase and elastase[5] in order to aid the process of remodeling and the decreases in fiber proteins that have been observed[61]. Perhaps the mononuclear cells so evident at some stages of regression[84] are also of importance as scavengers of lipids present in the necrotic center[85].

SUMMARY

This is a brief consideration of some of the most recent
and most significant developments that have helped us to understand the process of atherogenesis and the process of regression
of the nation's most lethal disease, namely atherosclerosis.

Beginning with a brief consideration of the most important
atherosclerotic lesion components and their relationship to
progression and regression, the major pathological developments
that have helped in understanding the initiation and pathogenesis
of the process are summarized. Some of the factors responsible
for the variability of the "fatty streak" and for the variability
of components of the developing human lesions are considered
briefly.

It appears that if sufficient lipid lowering is induced then
many components of the lesions may undergo regression with a
definite decrease in the size of the plaque. The components
that take part include a substantial decrease in the amount of
lipid in the plaque, the intracellular lipid and that between
cells, along with that which is pooled in the necrotic center.
There is also likely to be healing of the endothelial damage
over the plaque and some remodeling and perhaps even a decrease
in the fiber proteins of the plaque.

Although the best evidence of plaque regression now is
furnished by several studies in the rhesus model of advanced
atherosclerosis, the model which we believe most nearly resembles the human disease, we may be nearing a new era in
therapeutic studies. The time appears to be approaching when
it will be possible to evaluate therapy accurately by observing
the effect on the size of the plaques sequentially in relatively small groups of patients, perhaps using computer assisted,
non-invasive techniques.

Although some of the fundamental biological mechanisms by
which regression may take place are identifiable, much more work
will be necessary before plaque regression becomes established
as a well understood pathobiological process.

ACKNOWLEDGEMENTS

The author wishes to acknowledge the valuable contributions
of Dr. Dragoslava Vesselinovitch and many other colleagues,
not only to the preparation of this review, but also to much of
the work on regression in this laboratory. He also is grateful
to Mrs. Gertrud Friedman and Ms. Lauren Brown and Ms. Gwendolyn
Matthews for their assistance in developing this manuscript.

        The results of studies from this laboratory that are
summarized in this review have been largely supported by the
U.S. Public Health Service grants HL 15062, HL 17648, HL 6894,
as well as the Heart Research Foundation, Inc. and The Louis A •
Block Fund for Basic Research and Advanced Study at the
University of Chicago.

REFERENCES

1. H.C. McGill, Jr., The lesion , in: "Atherosclerosis III."
        G. Schettler and A. Weizel, eds., p. 27, Springer Verlag,
        Berlin (1974).
2. R.W. Wissler, Development of the atherosclerotic plaque, in:
        "The Myocardium. Failure and Infarction," E. Braunwald,
        ed. p. 155, H.P. Publishing Co. Inc., New York (1974).
3. R.W. Wissler, Principles of the pathogenesis of athero-
        sclerosis, in: "Heart Disease. A Textbook of Cardio-
        vascular Medicine," E. Braunwald, ed., Vol. 2, p. 1221,
        W.B. Saunders Co., Philadelphia (1980).
4. G.S. Berenson, S.R. Srinivasan, B. Radhakrishnamurthy, and
        E.R. Dalferes, Jr., Mucopholysaccharide-lipoprotein com-
        plexes in atherosclerotic aorta, in: "Arterial Mesenchyme
        and Arteriosclerosis," W.D. Wagner and T.B. Clarkson, eds.,
        p. 141, Plenum Press, New York (1974).
5. T. Schaffner, K. Taylor, E.J. Bartucci, K. Fischer-Dzoga,
        J.H. Beeson, S. Glagov and R.W. Wissler, Arterial foam
        cells exhibit distinctive immunomorphologic and histo-
        chemical features of macrophages, Am. J. Path. 100:57
        (1980).
6. M.D. Haust, R.H. More, and H.Z. Movat, The role of smooth
        muscle cells in the fibrogenesis of arteriosclerosis,
        Am. J. Path. 37:377 (1960).
7. J.C. Geer, H.C. McGill, Jr., J.P. Strong, and R.L. Holman,
        Electron microscopy of human atherosclerotic lesions,
        Fed. Proc. 19:15 (1960).
8. R.W. Wissler, The arterial medial cell, smooth muscle or
        multifunctional mesenchyme? J. Atheros. Res. 8:201 (1968).
9. R. Ross, The smooth muscle cell. II. Growth of smooth
        muscle in culture and formation of elastic fibers,
        J. Cell Biol. 50:172 (1971).
10. R.W. Wissler, D. Vesselinovitch, and G.S. Getz, Abnor-
        malities of the arterial wall and its metabolism in
        atherogenesis, Progr. Cardiovasc. Dis. 18:341 (1976).
11. R. Ross and J. Glomset, The pathogenesis of atherosclerosis
        (Part 1), New Eng. J. Med. 295:369 (1976).
12. R. Ross and J. Glomset, The pathogenesis of atherosclerosis
        (Part 2), New Eng. J. Med. 295:420 (1976).
13. R.W. Wissler, Overview of problems of atherosclerosis,
        in: "Cerebrovascular Disease," P.Scheinberg, ed.,
        p. 59, Raven Press, New York (1976).

14. M.L. Tiell and M.B. Stemerman, Suppression of myointimal hy-
    perplasia in hypophysectomized rats, Circulation 54
    (Suppl. II): 138 (1976).
15. R.W. Wissler, H.A. McAllister, and D. Vesselinovitch, A
    histopathological study of the fatty streak in aortas and
    coronary arteries of young American military personnel,
    Am J. Path. 78:64a (1975).
16. E.B. Smith and R.S. Slater, Lipids and low-density lipo-
    proteins in intima in relation to its morphological
    characteristics, in: "Atherogenesis: Initiating Factors"
    (Ciba Symposium), p. 39, Excerpta Medica, Amsterdam
    (1974).
17. R.V. Panganamala, J.C. Geer, H.M. Sharma, and D.G. Cornwell,
    The gross and histologic appearance and the lipid com-
    position of normal intima and lesions from human coronary
    arteries and aorta, Atherosclerosis 20:93 (1974).
18. J.R.A. Mitchell and C.J. Schwartz, "Arterial Disease,"
    Blackwell, Oxford (1965).
19. R.W. Wissler, Atherosclerosis - Its pathogenesis in pre-
    spective, in: "Comparative Pathology of the Heart."
    (Adv. Cardiol. Vol 13) F. Homberger, ed., p. 10,
    S. Karger, Basel (1974).
20. D. Vesselinovitch, G.S. Getz, R.H. Hughes, and R.W. Wissler,
    Atherosclerosis in the rhesus monkey fed three food
    fats, Atherosclerosis 20:303 (1974).
21. R.W. Wissler, D. Vesselinovitch, J. Borensztajn, T.Schaffner,
    and R. Hughes, Effect of various dietary responses on
    progression of atherosclerosis in rhesus monkeys,
    Fed. Proc. 35:294 (1976).
22. D. Vesselinovitch, R.W. Wissler, T.J. Schaffner, and J.
    Borensztajn, The effects of various diets on athero-
    genesis in rhesus monkeys, Atherosclerosis 35:189 (1980).
23. L.N. Katz and J. Stamler, "Experimental Atherosclerosis,"
    Thomas, Springfield, Ill. (1953).
24. P. Constantinides, "Experimental Atherosclerosis,"
    Elsevier, Amsterdam (1965).
25. J.C. Roberts and R. Straus, "Comparative Atherosclerosis,"
    Hoeber, New York (1965).
26. R.W. Wissler and D. Vesselinovitch, Evaluation of animal
    models for the study of the pathogenesis of atherosclero-
    sis, in: "International Symposium: State of Prevention
    and Therapy in Human Atherosclerosis and in Animal
    Models," W.H. Hauss, R.W. Wissler and R. Lehmann, eds.,
    p. 13, Westdeutscher Verlag, Opladen, W. Germany (1978).
27. D. Vesselinovitch, Animal models of atherosclerosis,
    their contributions and pitfalls, Artery 5:193 (1979).
28. G. Schettler, Cardiovascular diseases during and after
    World War II: A comparison of the Federal Republic
    of Germany with other European countries, Prev. Med.
    8:581 (1979).
29. M.R. Malinow, Regression of atherosclerosis in non-human pri-

mates. An overview, in: "The Use of Non-human Primates Cardiovascular Diseases," S.S. Kalter, ed., p. 181, Univ. Texas Press, Austin, (1980).

30. A.E. Dorr, K. Gundersen, J.C. Schneider, Jr., T.W. Spencer, and W.B. Martin, Colestipol hydrochloride in hyper-cholesterolemic patients - effect on serum cholesterol and mortality, J. Chron. Dis. 31:5 (1978).

31. N.J. Stone, Cholesterol, colestipol and coronary heart disease, J. Chron. Dis. 31:1 (1978).

32. O. Turpeinen, Effect of cholesterol-lowering diet on mortality from coronary heart disease and other causes, Circulation 59:1 (1979).

33. P. Leren, The Oslo diet - heart study: Eleven-year report. Circulation 42:935 (1970).

34. S.L. Wilens, The resorption of arterial atheromatous deposits in wasting disease, Am. J.Path. 23:793 (1947).

35. H. Buchwald, K. Amplatz, L. Knight, I. Guzman, and R.L. Varco, Arteriography changes after partial ileal bypass, in: "International Symposium: State of Prevention and Therapy in Human Arteriosclerosis and in Animal Models," W.H. Hauss, R.W. Wissler and R. Lehmann, eds., p. 469, Westdeutscher Verlag, Opladen, West Germany (1978).

36. R. Barndt, Jr., D.H. Blankenhorn, D.W. Crawford, and S.H. Brooks, Regression and progression of early femoral atherosclerosis in treated hyperlipoproteinemic patients, Ann. Int. Med. 86:139 (1977).

37. I. Vartiainen and K. Kanerva, Artherosclerosis in wartime, Ann med. int. Fenniae 36:748 (1947).

38. K. Solth, R. Kohl, G. Schettler, and A. Werthemann, Zur Statistik der Arteriosklerose, Verh. Deut. Ges. Pathol. 41:64: (1957).

39. U.A. Rivin and S.P. Dimitroff, The incidence and severity of atherosclerosis in estrogen-treated males, and in females with a hypoestrogenic or a hyperestrogenic state, Circulation 9:533(1954).

40. M.S. Moskowitz, A.A. Moskowitz, W.L. Bradford, and R.W. Wissler, Changes in serum lipids and coronary arteries of the rat in response to estrogens, Arch. Path. 61:245 (1956).

41. J. Stamler, R. Pick, and L.N. Katz, Experiences in assessing estrogen antiatherogenesis in the chick, the rabbit and man, Ann. N.Y. Acad. Sci. 64:596 (1956).

42. W.T. London, S.E. Rosenberg, J.W. Draper, and T.P. Almy, The effect of estrogen on atherosclerosis, Ann. Int. Med. 55:63 (1961).

43. H. Buchwald, R. B. Moore and R.L. Varco, Surgical treatment of hyperlipidemia, Circulation 49 (suppl. I.):1 (1974).

44. P. Constantinides, J. Booth, and G. Carlson, Production of advanced cholesterol atherosclerosis in the rabbit, Arch. Path. 70:712 (1960).

45. R.W. Wissler, and D. Vesselinovitch, Experimental models of human atherosclerosis, Ann. N.Y. Acad. Sci. 149:907 (1968).

46. R.W. Wissler, and D. Vesselinovitch, Difference between human and animal atherosclerosis, in: "Atherosclerosis III," G. Schettler and A. Weizel, eds., p. 319, Springer-Verlag, Berlin (1974).

47. R. Wissler and D. Vesselinovitch, Studies of regression of advanced atherosclerosis in experimental animals and man, in: "Atherogenesis," Ann. N.Y. Acad. Sci. 275: 363 (1976).

48. D. Vesselinovitch and R.W. Wissler, Comparison of primates and rabbits as animal models in experimental athero-sclerosis, in: "Atherosclerosis: Metabolic, Morphologic and Clinical Aspects," G.W. Manning and M.D. Haust; eds., p. 614, Plenum Press, New York (1977).

49. R.W. Wissler, and D. Vesselinovitch, Animal models of re-gression, in: "Atherosclerosis IV," G. Schettler, Y. Goto, Y. Hata and G. Klose, eds., p. 377, Springer-Verlag, Berlin (1977).

50. R.W. Wissler and D. Vesselinovitch, Evaluation of animal models for the study of the pathogenesis of athero-sclerosis, in: "International Symposium. State of Pre-vention and Therapy in Human Arteriosclerosis and in Animal Models," W.H. Hauss, R.W. Wissler and R. Lehmann, eds., p. 13, Westdeutscher Verlag, Opladen, West Germany (1978).

51. D. Vesselinovitch and R.W. Wissler, Prevention and regres-sion in animal models by diet and cholestyramine, in: "International Symposium. State of Prevention and Therapy in Human Arteriosclerosis and in Animal Models," W.H. Hauss, R.W. Wissler, and R. Lehmann, eds., p. 127, Westdeutscher Verlag, Opladen, West Germany (1978).

52. A. Wartman, T.L. Lampe, P.S. McCann, and A.J. Boyle, Plaque reversal with Mg EDTA in experimental athero-sclerosis: elastin and collagen metabolism, J. Athero-scler. Res. 7:331 (1967).

53. K. Kjeldsen, P. Astrup, and J. Wanstrup, Reversal of rab-bit atheromatosis by hyperoxia, J. Atheroscler. Res. 10:173 (1969).

54. D. Vesselinovitch, R.W. Wissler, K. Fischer-Dzoga, R. Hughes, and L. DuBien, Regression of atherosclerosis in rabbits. I. Treatment with low fat diet, hyperoxia and hypolipi-demic agents, Atherosclerosis 19:259 (1974).

55. C.B. Taylor, G.E. Cox, P. Manalo-Estrella, and J. South-worth, Atherosclerosis in rhesus monkeys. II. Arterial lesions associated with hypercholesterolemia induced by dietary fat and cholesterol, Arch. Pathol. 74:16 (1962).

56. C.B. Taylor, D.E. Patton, and G.E. Cox, Atherosclerosis in rhesus monkeys. VI. Fatal myocardial infarction in a monkey

fed fat and cholesterol. Arch. Pathol. 76:404 (1963).

57. C.B. Taylor, Experimentally induced atherosclerosis in non-
     human primates, in: "Comparative Atherosclerosis,"
     J.C. Roberts and R. Strauss, eds., p. 215, Hoeber,
     New York (1965).

58. M.L. Armstrong, E.D. Warner, and W.E. Connor, Regression
     of coronary atheromatosis in rhesus monkeys, Circul.
     Res. 27:59 (1970).

59. R.W. Wissler, Current status of regression studies, in:
     "Atherosclerosis Reviews," Vol. 3, R. Paoletti and
     A.M. Gotto, eds., p. 213, Raven Press, New York (1978).

60. M.L. Armstrong and M.B. Megan, Lipid depletion in athero-
     matous coronary arteries in rhesus monkeys after regres-
     sion diets, Circul. Res. 30:675 (1972).

61. M.L. Armstrong, Connective tissue changes in regression, in:
     "Atherosclerosis IV," G. Schettler, Y. Goto, Y. Hata,
     G. Klose, eds., p. 405, Springer-Verlag, Berlin (1977).

62. D. Vesselinovitch, R.W. Wissler, R. Hughes, and J. Borensztajn,
     Reversal of advanced atherosclerosis in rhesus monkeys.
     I. Gross and light microscopic studies, Atherosclerosis
     23:155 (1976).

63. R.W. Wissler, unpublished observation.

64. R.M. Jones, and R.W. Wissler, Surface features of experi-
     mental atherosclerotic plaques in rhesus monkeys with
     and without treatment, in: Scanning Electron Microscopy,
     Part II., O. Johari, ed., p. 975, SEM, Inc., Chicago
     (1978).

65. G. Weber, P. Fabbrini, L. Resi, R. Jones, D. Vesselinovitch, and
     R.W. Wissler, Regression of arteriosclerotic lesions
     in rhesus monkey aortas after regression diet: scanning
     and transmission electron microscope observations of the
     endothelium, Atherosclerosis 26:535 (1977).

66. J. Borensztajn, K. Foreman, R.W. Wissler, H. VanZutphen,
     D. Vesselinovitch, and R. Hughes, Egress of aortic cho-
     lesterol and cholesterol ester during regression of athero-
     sclerosis in rhesus monkeys, Circulation 52 (Suppl. II):
     269 (1975).

67. D. Vesselinovitch, R.W. Wissler, J. Borensztajn, and T.J.
     Schaffner, The effects of diets with and without cho-
     lestyramine on the lesion compoments of atherosclerotic
     plaques, Fed. Proc. 37:835 (1978).

68. D. Vesselinovitch, R.W. Wissler, and T.J. Schaffner, Quantita-
     tion of lesions during progression and regression of
     atherosclerosis in rhesus monkeys. Symposium on Cardio-
     vascular Disease in Nutrition (19th Annual Meeting, Am.
     Assn. Nutrition, U. Minn.), Spectrum Publications, New
     York, in press (1979).

69. R.W. Wissler, D. Vesselinovitch, T.J. Schaffner, and S.
     Glagov, Quantitating rhesus monkey atherosclerosis pro-
     gression and regression with time, in: "Atherosclerosis V,"

A.M. Gotto, Jr., L.C. Smith and B. Allen, eds., p. 757, Springer-Verlag, New York (1980).

70. D.A. Eggen, J.P. Strong, and H.C. McGill, Jr., An objective method for grading atherosclerotic lesions, Lab. Invest. 11:732 (1962).

71. C.F. Howard, Jr., Aortic atherosclerosis in normal and spontaneously diabetic Macaca nigra, Atherosclerosis 33:479 (1979).

72. B. Radhakrishnamurthy, H.A. Ruiz, E.R. Dalferes, Jr., D. Vesselinovitch, R.W. Wissler and G.S. Berenson, The effect of various dietary regimens and cholestyramine on aortic glycosaminoglycans during regression of atherosclerotic lesions in rhesus monkeys, Atherosclerosis 33:17 (1979).

73. A.J. Daoud, J. Jarmolych, J. Augustyn, K.E. Fritz, J.K. Singh, and K.T. Lee, Regression of advanced swine atherosclerosis, Arch. Pathol. Lab. Med. 100:372 (1976).

74. K.E. Fritz, J. Augustyn, J. Jarmolych, A.S. Daoud and K.T. Lee, Regression of advanced atherosclerosis in swine (chemical studies), Arch. Pathol. Lab. Med. 100:380 (1976).

75. R.H. Knopp, Test of the lipid hypothesis: The coronary primary prevention trial (COPT) of the lipid research clinics program, in:"Atherosclerosis V," A.M. Gotto, Jr., L.C. Smith, B. Allen, eds., p. 509, Springer-Verlag, New York (1980).

76. D.W. Crawford, E.S. Beckenbach, D.H. Blankenhorn, R.H. Selzer, and S.H. Brooks, Grading of coronary atherosclerosis. Comparison of a modified IAP visual grading method and a new quantitative angiographic technique, Atherosclerosis, 19:231 (1974).

77. D.W. Crawford and D.H. Blankenhorn, Quantitative studies of atherosclerotic plaque size in human subjects, in: "International Symposium: State of Prevention and Therapy in Human Arteriosclerosis and in Animal Models," W.H. Hauss, R.W. Wissler, and R. Lehmann, eds., p. 55, Westdeutscher Verlag, Opladen, West Germany (1978).

78. R.W. Barnes, Doppler ultrasonic arteriography and flow velocity analysis in carotid artery disease, in:"Noninvasive Diagnostic Techniques in Vascular Disease," E.F. Borenstein, ed., p. 221, C.V. Mosby Company, St. Louis (1978).

79. D. Nash, Personal communication.

80. D.H. Blankenhorn, Personal communication.

81. R. Ross and J.A. Glomset, The pathogenesis of atherosclerosis, New Eng. J. Med. 295:369 (1976).

82. C.M. Gajdusek, P. DiCorleto, R. Ross, and S.M. Schwartz, An endothelial cell derived growth factor, J. Cell Biology 85:467 (1980).

83. K. Fischer-Dzoga, R. Fraser, and R.W. Wissler, Stimulation of proliferation in stationary primary cultures of monkey and rabbit aortic smooth muscle cells. I. Effects of lipoprotein fractions of hyperlipemic serum and lymph, Exp. Mol. Pathol. 24:346 (1976).

84. H.C. Stary, Ultrastructural changes in the lipid inclusions of a arterial smooth muscle cells after reduction of high serum cholesterol levels, Progress in Biochem. Pharmacol 14:46 (1977).

85. A.S. Daoud, J. Jarmolych, J.M. Augustyn, and K.E. Fritz, Sequential morphologic studies of regression of advanced atherosclerosis, Arch. Path. 105:233 (1981).

ULTRASOUND TECHNIQUES IN THE STUDY OF CAROTID ARTERY ATHEROMA

Robert J. Lusby

University Dept. Surgery, Bristol Royal Infirmary
Bristol, England
St. Vincent's Hospital, Sydney, Australia

INTRODUCTION

Cerebrovascular disease is the third most frequent cause of
death in western society. In Europe over one million new cases
of stroke are reported each year, with a 30 percent mortality and
a 60 percent incidence of substantial disability among the
survivors. In 1973 it was estimated that over 7 billion dollars
was spent annually in the initial hospital care of these patients
(1,2). It is generally estimated that 60-70 percent of strokes
are due to thromboembolism (3,4). The importance of extracranial
atherosclerotic disease in the production of stroke has been
recognised only in the last 30 years (5,6). Recent studies have
shown up to 88% of patients with cerebral ischaemia to have
extracranial artery lesions (7,8) with the majority situated at
the carotid artery bifurcation (9). There are two basic
mechanisms by which carotid artery lesions produce cerebral
ischaemia (A) reduced flow from a high grade stenosis or occlusion
and (B) from emboli generated from the region of atheromatous
plaques.

Stroke from extracranial atherosclerosis should not be seen
as an isolated event - "a cerebral vascular accident", but rather
as the end stage of a disease that has progressed from the first
appearance of the atherosclerotic plaque through early clinical
manifestations such as asymptomatic carotid bruits, transient
cerebral ischaemic episodes to the final fixed neurological defeat
or death.

As there is no known treatment to limit or reverse cerebral infarction once it has started, effective therapy must depend on early recognition and treatment of extracranial lesions. Fortunately not all strokes come without warning and there exist a group of prestroke syndromes - transient ischaemic attacks (TIAs), Amaurosis fugax, and asymptomatic bruits which may be treated in an effort to prevent stroke. However, approximately half the patients with TIAs and two-thirds with asymptomatic bruits will not proceed to stroke if left untreated. The recognition of the significant predeterminants of stroke is a difficult clinical problem which is of importance in the selection of appropriate medical therapy.

Recent developments in ultrasound techniques have made it possible to non invasively study patients with arterial disease using Doppler blood velocity signal analysis (10-12) and direct imaging of atheromatous lesions at specific sites (13-18). We report the use of these techniques as a means of detecting carotid artery lesions and defining some of the factors that contribute to cerebral infarction.

Materials and Methods

One hundred and eighty-nine patients with prestroke syndromes have been studied in the vascular laboratory of the Bristol Royal Infirmary since January 1979. Two ultrasonic imaging systems have been used.

Pulsed Doppler System

MAVIS, a 30-channel, range-gated, 5 MHz pulsed Doppler system, was developed by Fish (13). The 30 gates can be adjusted to a suitable depth and the movement of blood detected by the Doppler shift principle in any or all of the gates. By moving the probe manually across the skin an image of the moving column of blood is built up on a storage oscilloscope, a permanent record of which is obtained by means of a polaroid photograph. The machine is directional, producing separate images of arterial blood moving towards the beam and venous blood moving away. Images can be produced in three orthogonal planes (Fig.1). The resolution is of the order of 1 mm.

Real-time B-mode Imaging

The Duplex Real-time mechanical scanner produces a two-dimensional B-mode image from three 5 MHz transducers which rotate in a plastic boot. The images obtained in longitudinal or cross-sectional planes are displayed on an oscilloscope. Superimposed on the display is a line corresponding to the location of a depth-sensitive pulsed 5 MHz ultrasound beam (Fig.2) which can be

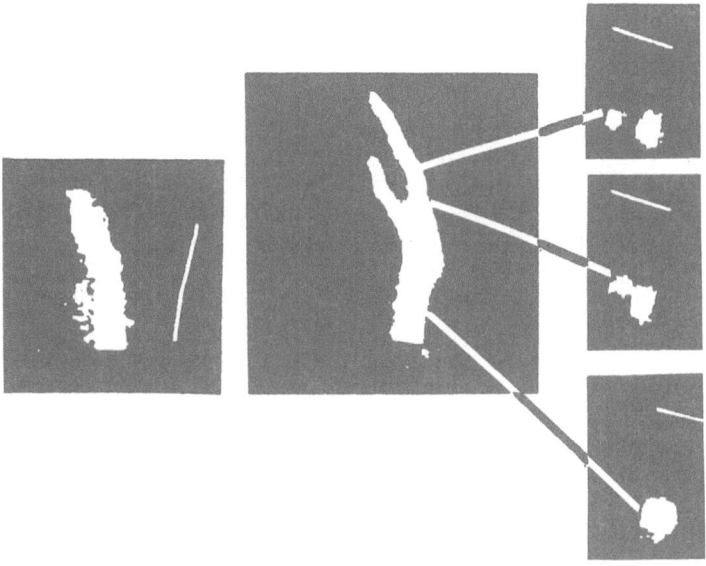

FIG. 1 MAVIS pulsed Doppler projections in lateral,
anteroposterior and cross-sectional planes.

adjusted to provide signals from selected sites for analysis. A
photocopy of a frozen image is obtained for the permanent record.
The Duplex scanner also has the facility for generating time
position M-mode scans. In this mode echoes are recorded from
moving structures such as the vessel walls and displayed as
vertical deflections (19). The vessel wall distensibility with
each pulse can therefore be displayed dynamically on the screen.
By adding a scale, changes in vessel diameter with each pulse can
be quantified. Using the B-mode scan the vessel to be studied
was clearly identified and distensibility measured in the common
carotid and origin of the internal carotid arteries.

To study the dynamic properties of the carotid bifurcation
further biplanar video tape recordings and cineangiography films
at 50 frames/sec were made during the injection of contrast into
12 carotid arteries of patients undergoing investigation for TIAs.
To ensure complete dye mixing with the blood and limit boundary
layer separation the catheter tip was placed in the proximal common
carotid artery near its origin from the aortic arch.

To evaluate the morphology of the carotid bifurcation and its
relation to ultrasonic and angiographic images barium impregnated

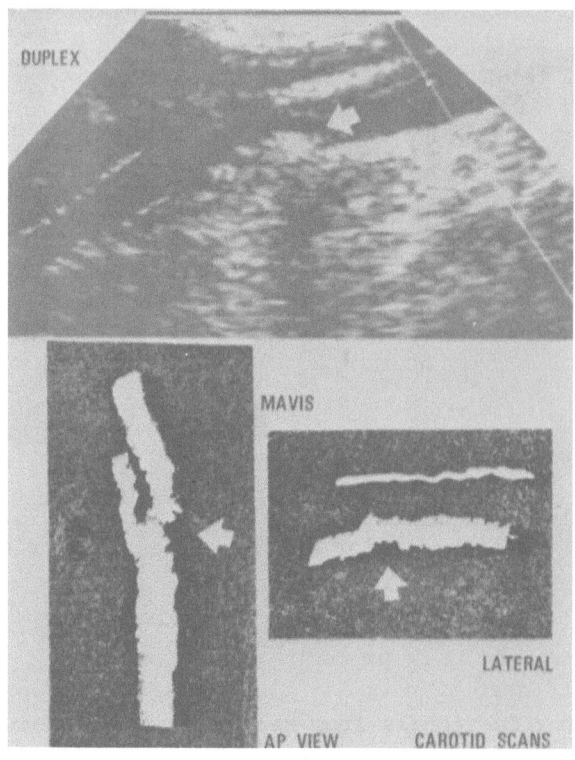

FIG. 2 Duplex Scan above and MAVIS anteroposterior
        and lateral scan below of the same lesion, arrowed.
        The white line on the left of the Duplex Scan
        represents the pulsed doppler beam and the dot
        the place of the sample volume.

perspex casts were made from the endarterectomy specimens of
patients undergoing carotid surgery and from vessels obtained at
post mortem.

Results

Ultrasonic Imaging

Both imaging systems had a high degree of patient acceptability
providing images of the vessels without discomfort or complication.
Seventy-one patients subsequently underwent angiography, at the
discretion of the referring physician, with corroboration of the
ultrasonic findings in 68 (90%). A detailed prospective evaluation
of the systems was performed during the study period on 78 vessels
in which three planar angiography was performed, each modality
being independently reported on without knowledge of the other
findings (20,21).

Pulsed Doppler Imaging

The comparison between MAVIS imaging and angiography is given
in Table 1. Of the 43 lesions of less than 50% diameter reduction
39 (90.6%) were detected, there were four false negatives (lesions
detected angiographically but missed on ultrasound imaging),
two were associated with atheromatous plaques in bulbous internal
carotid origins. The lateral scan was better at detecting lesions
lying on the posterior wall and provided the diagnostic information
in 20 vessels (48%) with less than 50% diameter reduction. All
five occlusions of the internal carotid artery were detected.

TABLE 1. MAVIS pulsed Doppler carotid bifurcation assessment

Angiographic findings	No. of Vessels	Positive No.	%	Negative	Positive other* estimate
Normal	16	1		94%	
Less than 25%	26	22	85	4%	
25 - 50%	17	14	82		3 under
50 - 99%	14	12	86		2 under
Occluded	5	5	100		
	78				

* where a lesion was detected but the estimate of severity
  did not agree with that seen on angiography.

Duplex B-Mode Imaging

The comparison between Duplex imaging and angiography is given in Table 2. While 39 (90.6%) of lesions were detected, four were given a classification at variance with that seen on angiography. All lesions of greater than 50% diameter reduction were detected but only four out of five of those that were totally occluded.

The essential difference between the two imaging systems is seen in Figure 3. where, in the lateral MAVIS scan, the lesion is seen as a defect in the blood flow map while the Duplex scan shows a vessel wall plaque encroaching on the lumen.

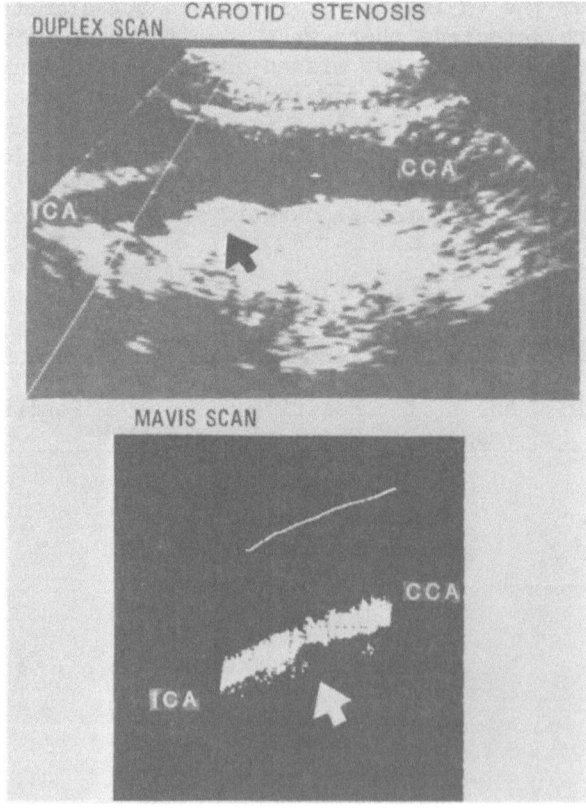

FIG.3. Lateral Duplex and MAVIS scans showing a lesion, arro-
        wed. The Duplex scan shows encroachment into the lumen
        by the lesion while the Mavis scan shows a defect in
        the image of the moving column of blood.

TABLE 2.   Duplex B mode carotid bifurcation assessment

Angiographic findings	No. of Vessels	Positive No.	%	Negative	Positive other* estimate
Normal	16	1		94%	
Less than 25%	26	19	73	4%	3
25 - 50%	17	16	94		1
50 - 99%	14	13	93		1
Occluded	5	4	50	1	
	78				

* where a lesion was detected but the estimate of severity
  did not agree with that seen on angiography.

The ultrasonic classification of lesions was made solely on
imaging although in producing the image the operator, with
experience, was guided by the presence of turbulent flow in the
Doppler audio signal to look further for a lesion if not at
first obvious. Several images were made of each vessel to confirm
a persistent defect. The overall sensitivity in detecting all grades
of lesion was 95% for the MAVIS pulsed Doppler and 93% for the
Duplex system. The specificity for both was 94%.

Turbulence near the site of an atheromatous plaque was
commonly noted and an example of this is shown in Figure 4. Here
the profile of the blood velocities in the common carotid artery
proximal to a lesion is relatively normal. Distal to the stenosis
there is an increase in the maximum velocity and a broader spectrum
of velocities arising from the turbulence and disorganisation of the
normal laminar flow.

Using the Duplex system it was necessary to carefully insonate
the vessels and confirm the presence of arterial flow. Soft
thrombus in five patients completely occluded the internal carotid
artery but, due to the low acoustic impedance of thrombus which is
identical to fluid blood, the B-mode image failed to identify the
occlusion. Failure to obtain a Doppler signal in four patients
when the sample volume was placed in the vessel lumen indicated
the absence of flowing blood even though the image appeared
relatively normal. In the fifth, a patent external carotid artery
was mistaken for the internal carotid artery.

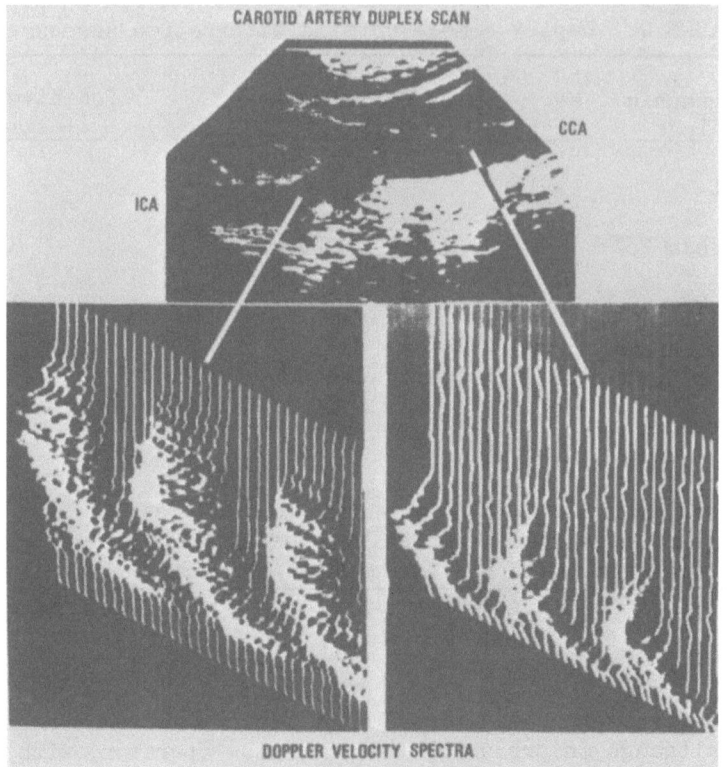

FIG. 4   Doppler velocity spectral analysis proximal and
         distal to a small lesion in the carotid bulb.
         There is a turbulent pattern with increased
         velocity and spectral broadening shown just
         distal to the lesion

Vessel distensibility studies

Using the Real-time Duplex system it was noticed that the
luminal diameter varied with the systolic to diastolic changes in
vessel wall distension. Dynamic recordings of contrast injection
during angiography in 12 vessels have shown the average variation
in internal diameter of the common carotid artery to be mean
±SEM 16.2 ±8.5 per cent, while at the bulbous origin of the internal
carotid artery the average variation in diameter was means ± SEM
3.4 ± 15.6 ($\underline{P}$ < 0.005 Wilcoxon)(Figure 5).

The time position M-mode scans performed in 10 patients
showed the average distensibility of the common carotid artery to
be mean ± SEM 12.6 ± 2.4, while in the bulbous origin of the
internal carotid arteries this was mean ± SEM 19.7 ± 5.4 per cent
($\underline{P}$ < 0.005)(Figure 6).

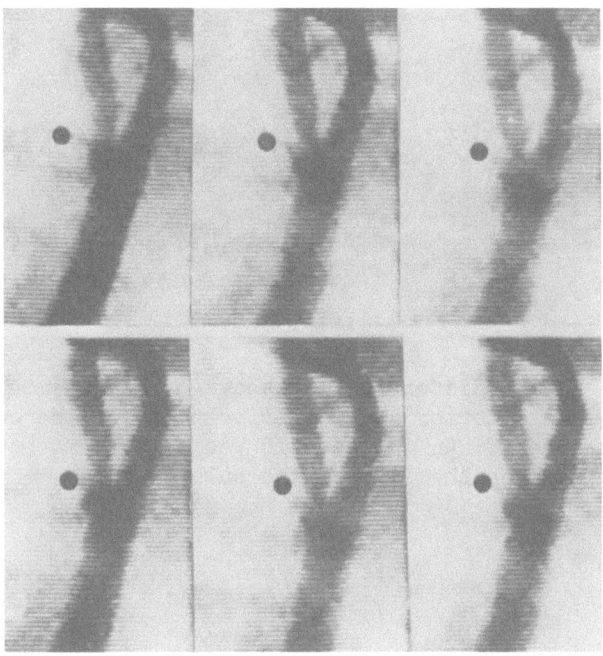

FIG. 5   Six video frames from a carotid angiogram.
         There is a marked variation in the internal
         diameter at the origin of the internal
         carotid artery (black dot).

The internal diameter of the bulb compared to that of the
distal internal carotid artery on single shot angiography was
measured in 23 vessels with bulbous origins.   The average bulb was
157% (range 120 - 246%) that of the distal vessel.(22)

Correlation of Ultrasonic Findings with Clinical Data

One hundred and eighty-nine patients were referred for study to
the vascular laboratory with symptoms suggesting episodes of transient cerebral ischaemia or Amaurosis fugax.   Eighty-six percent of
patients had a vessel wall abnormality detected by ultrasound
imaging.   Among the 16% without evidence of carotid bifurcation
disease were patients with known valvular heart disease, cardiac

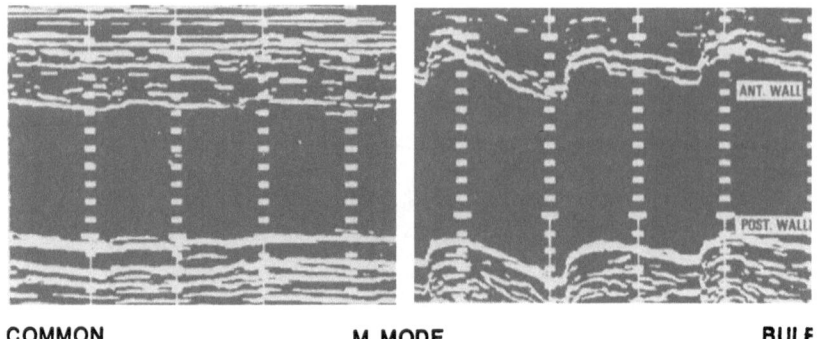

COMMON   M MODE   BULB

FIG 6   M Mode distensibility showing the movement in
        the anterior and posterior walls of the carotid
        artery. An increase in vessel wall movement is
        seen on the side of the bulb.

arrhythmias or evidence of subclavian or vertebral artery disease.

    Forty-five percent of patients with hemisphere transient
ischaemic attacks or Amaurosis fugax had bruits in the neck over the
appropriate carotid artery. Fifty-four percent of bruits were
associated with lesions greater than 50%, while 46% were associated
with lesser lesions.

    Of symptomatic patients 40% had ultrasound detected lesions of
greater than 50% diameter reduction, the remainder having lesions
with less than 50% diameter reduction. There was no difference
between the size of lesion and the manifestation of either hemisphere TIAs or Amaurosis fugax with 45 and 47% of lesions being
greater than 50% stenosis respectively.(22)

Discussion

    The ability to detect all grades of lesions is essential if a
non invasive test is to be applied to screening for carotid artery
disease. In this study both ultrasonic imaging systems demonstrated
their capability in the detection of all grades of stenosis. The
advantage of the pulsed Doppler system with range-gating is that it
can produce images in three orthogonal directions, thus providing
three dimensional information about the internal lumen of the vessel.
In particular this study shows how valuable the lateral scan is in

detecting lesions, where in many instances the antero-posterior view
failed to show a lesion. The nature of carotid artery atheroma to
be deposited on the posterior wall makes it necessary to carefully
inspect this area, the Duplex system readily provides views of this
wall, as does the lateral MAVIS scan.

The first report of pulsed Doppler imaging using a 6 channel
system (14) showed an overall agreement to within one grade of 70%
and as a sensitivity of 90% but a low specificity of 44%. However
this system does not provide lateral scans and thus reduces the
ability to detect small lesions. Sumner et al (1979), (17) using
the same equipment, excluded lesions of less than 20% stenosis to
achieve a sensitivity of 86% and specificity of 90%. These minor
lesions can be associated with symptoms and the MAVIS 30 channel
system with its multiplanar imaging appears to be more sensitive
in the detection of all lesions (93%).

The imaging capability of the Duplex scanner cannot be separated from the pulsed Doppler as identification of the type of
vessel (vein or artery) is provided by the Doppler. In addition,
the presence of fresh thrombus occluding a vessel can only be
detected with the Doppler sampling.

Disturbances of flow in the region of arterial stenosis using
Doppler techniques were first described by Strandness (1967) (23)
and these changes have been quantified using Doppler waveform analysis (10-12). Blackshear et al. using the Duplex system have
quantified the spectral broadening produced by turbulence near
carotid lesions and were able to distinguish between normal, stenotic
and occluded arteries on the basis of the velocity pattern alone
and noted differences in the velocity spectrum between high and low
grade lesions.

The carotid artery bruit represents relatively high frequency
vessel wall movement in the audible range (24). Within this range
there is evidence of specific frequencies for individual vessels
causing changes in collagen and elastin structural components
(24,25). With these changes can come vessel wall dilation,
ulceration and exposure of wall components which are known platelet
activating agents, e.g. collagen and basement membrane.

The pulsatile changes in vessel wall diameter appear much
greater than previously appreciated. Studies of exposed intact
vessels have shown changes of 1 to 2% (27), whereas in this study
both M-mode and dynamic angiographic findings have confirmed the
earlier report by Arndt (28), using an echo tracking technique,
of increased in vivo distensibility.

In particular, in our study a previously unappreciated
increased distensibility has been demonstrated in the carotid

artery bulb, even in the presence of atheromatous involvement. Pathological studies have suggested the possibility of increased distensibility with the bulb having a reduced medial thickness relative to the vessel wall on either side, giving it a structure more appropriate to the pulmonary vessels (29).

The variation in distensibility seen probably represents abrupt segmental changes in vessel compliance, particularly in the region of the internal carotid artery origin. The stress associated with such abrupt changes in compliance has been associated with disruptive vessel wall forces resulting in, for example, aneurysm formation and thrombus deposition (30-32). The same forces may well result in plaque fracture which has recently been re-emphasised as a cause of platelet aggregation and thrombosis.

Turbulence and disorganisation of normal laminar flow distal to a stenotic lesion, as detected with the pulsed Doppler and characterised by spectral analysis, may also be important de novo in the initiation of platelet aggregation and thromboembolism (33-34). Such turbulence is invariably present in lesions of a magnitude sufficient to produce a drop in pressure and flow to the cerebral hemisphere (11). These haemodynamically significant lesions have been demonstrated to frequently result in cerebral infarction and stroke (35). Increased velocity gradients and jet streaming effects may also cause vascular endothelial (36) damage leading to platelet aggregation.

These preliminary results show that ultrasound techniques may be applied to the study of carotid atheroma. Some of the factors that lead to the generation of emboli may be mechanical in nature and further investigation using non-invasive techniques may help define the events that contribute to stroke.

REFERENCES

1. Haberman, S., Capildeo, R., and Clifford Rose, F., Epidemiological aspects of stroke, In: Progress in Stroke Research 1, R. Greenhalgh and F. Clifford Rose, eds., Pitman Medical, pp. 3-14 (1979).

2. Jurgen, M., World Health Organization Stroke Register, Medical Tribune (1973).

3. Kannel, W.B., Wolf, P.A., Verter, J., et al., Epidemiologic assessment of the role of blood pressure and stroke - The Framingham Study, J.A.M.A. 214: 301-310 (1970).

4. Whisnant, J.P., Matsumoto, N., and Elvebeck L.R., Transient cerebral ischaemic attacks in a community, Mayo Clin.Proc. 48: 194-198 (1973).

5. Fisher, M., Occlusion of the internal carotid artery, Arch. Neurol. Psychiat. 72: 182-204 (1951).

6. Yates, P.O., and Hutchinson, E.C. Cerebral infarction. The role of stenosis of the extracranial cerebral arteries. In: Medical Research Council Special Report Series No. 300, HMSO, London (1961).

7. Hutchinson, Acheson, Strokes: Natural History, Pathology and Surgical Treatment, W.B. Saunders, London (1975).

8. Eisenberg, R.L., Nemsek, W.R., Moore, W.S., and Mani, R.L., Relationship of transient ischaemic attacks and angiographically demonstrable lesions of carotid artery, Stroke 8: 483 (1977).

9. Hass, W.K., Fields W.S., North, R.R., Kricheff, I.I., Chase, N.E., and Bauer, B.B., Joint study of extracranial arterial occlusion: (1) Arteriography, techniques, sites and complications, J.A.M.A. 203: 961-968 (1968).

10. Woodcock, J.P., King, D.H., Gosling, R.G., and Nueman, D.L., Physical aspects of the measurement of blood velocity by Doppler shifted ultrasound, In: Blood Flow Measurement, C. Roberts, ed., Chapter 1, Sector Publ., London (1972).

11. Blackshear, W.M., Phillips, D.J., Thiele, B.L., Birsch, J.H., Chikos, P.M., Marinelli, M.R., Ward, K.J., and Strandness D.E., Detection of carotid occlusive disease by ultrasonic imaging and pulsed Doppler spectral analysis, Surgery 86: 698-707 (1979).

12. Baird, R.N., Bird, D.R., Clifford, P.C., Lusby, R.J., Skidmore, R., and Woodcock, U.P., Upstread stenosis: Its diagnosis by Doppler signals from the femoral artery, Arch.Surg. 115: 1316-1322 (1980).

13. Fish, P.J. In: Blood Flow Measurement, V.S. Roberts, ed., Sector Publ. p.29 (1972).

14. Mozersky, D.J., Bauer, D.W., Hokanson, D.E., Sumner, D.S., and Strandness, D.E., Ultrasonic arteriography, Arch.Surg. 103: 663-667 (1971).

15. Barnes, R.W., Bone, G.E., Reinerston, J.E., Slaymaker, E.E., Hokanson, D.E., and Strandness, D.E. Jr., Noninvasive ultrasonic carotid angiography. Prospective validation by contrast arteriography, Surgery 80: 328 (1976).

16. Baird, R.N., Lusby, R.J., Bird D.R., Giddings, A.E.B., Skidmore, R., Woodcock, U.P., Horton, R.E., and Peacock, J.H., Pulsed Doppler angbiography in lower limb ischaemia, Surgery 80: 818-825 (1979).

17. Lusby, R.J., Pulsed Doppler assessment of the profunda femoris artery, In: Diagnosis and Monitoring in Arterial Surgery, R.N. Baird and J.P. Woodcock, eds., John Wright & Sons, Bristol (1980).

18. Sumner, D.S., Russell, J.B., Ramsey, D.E., Hajjar, W.M., and Miles, R.D., Noninvasive diagnosis of extracranial carotid arterial disease: A prospective evaluation of pulsed Doppler imaging and oculoplethysmography, Arch.Surg. 114: 1222-1229 (1979).

19. Lusby, R.J., Clifford, P.C., Bird, D.R., Skidmore, R., Woodcock, J.P., and Baird, R.N., Ultrasonic techniques in the study of graft survival, In: Haemodynamics of the limbs (11) P. Puel, ed., (in press).

20. Lusby, R.J., Woodcock, J., Skidmore, R., Jeans, W., Clifford, P.C. and Baird, R.N., Carotid artery disease: A prospective evaluation of ultrasonic imaging in the detection of low and high grade stenosis, Br.J.Surg. 67: 823 (1980).

21. Lusby, R.J., Woodcock, J.P., Skidmore, R., Jeans, W.D., Hope, D.T., and Baird, R.N., Carotid artery disease: A prospective evaluation of pulsed Doppler imaging ultrasound, Med. Biol. (in press).

22. Lusby, R.J., Machleder, H.I., Jeans, W., Skidmore, R., Woodcock, J.P., Clifford, P.C., and Baird, R.N., Characterisation of vessel wall and blood flow dynamics in arterial thromboembolism, Proc.Roy. Soc. (in press).

23. Strandness, D.E., Jr., Schultz, R.O., Sumner, D.S., Rushmer, R.F., Ultrasonic flow detection: A useful technique in the evaluation of peripheral vascular disease, Am.J.Surg. 113: 311-320 (1967).

24. McDonald, D.A., Blood Flow in Arteries (2nd ed.), Edward Arnold, London, and Williams and Wilkins, Baltimore (1974).

25. Gersten, J.W., Relation of ultrasound effects to orientation of tendon in ultrasound field, Arch. Phys.Med.Rehabil. 37: 201 (1956).

26. Boughner, D.R., and Roach, M.R., Effect of low frequency vibration on arterial wall, Circ.Res. 29: 136 (1971).

27. Greenfield, J.C., Tindall, G.T., Dillon, M.L., and Mahaley, M.S., Mechanics of the human carotid artery in vivo, Circ.Res. 15: 240 (1964).

28. Arndt, J.O., Klauske, J., and Mersch, F., The diameter of the intact carotid artery in man and its change with pulse pressure, Pfluegers Arch. 301: 230 (1968).

29. Heath, D., Smith, P., Harris, P., and Winson, M., The atherosclerotic human carotid sinus, J.Path. 110: 49-58 (1973).

30. Gonza, E.R., Mason, W.F., Marble, A.E., Winter, D.A., and Dolan, F.G., Necessity for elastic properties in synthetic arterial grafts, Can.J.Surg. 17: 1-5 (1974).

31. Eiken, O. Pressure-flow relationship and thrombotic occlusion of experimental grafts, Acta Chir.Scand. 121: 398 (1961).

32. Baird, R.N., and Abbott, W.M., Pulsatile blood flow in arterial grafts, Lancet ii: 948-950 (1976).

33. Mustard, J.F., Murphy, E.A., Rowsell, H.C., and Downie, H.G., Factors influencing thrombus formation in vivo, Am.J.Med. 33: 621-646 (1962).

34. Born, G.V.R., Arterial thrombosis and its prevention, S. Hayase and S. Murao, eds., Proc. VIII World Congress of Cardiology, Tokyo, Excerpta Medica, Amsterdam (1978).

35. Machleder, H.I., Strokes, transient ischaemic attacks and asymptomatic bruits, West.J.Med. 130: 205-217 (1979).

36. Fry, D.L., Acute vascular endothelial changes associated with increased velocity gradients, Circ. Res. 22: 165 (1968).

EXPERIMENTAL ATHEROSCLEROSIS: DIET AND DRUGS

David Kritchevsky

The Wistar Institute of Anatomy and Biology
36th and Spruce Streets
Philadelphia, Pennsylvania 19104, U.S.A.

The earliest studies of induced atherosclerosis involved the feeding of milk and eggs (1) or of pure cholesterol to rabbits (2). The choice of the rabbit for this type of study has been criticized on the grounds that this species is normally herbiverous and thus we are studying a cholesterol storage disease in an animal that cannot detoxify excess cholesterol. The criticism is justifiable but, in fact, any animal species used differs metabolically from man to some extent. While direct extrapolation of animal data to man is generally unwarranted these experiments provide clues which may be useful in assessing dietary or drug treatment.

The induction of atherosclerosis in animal models can be achieved in a number of ways. The most commonly used experimental model is the rabbit. In the rabbit, atherosclerosis is established by feeding cholesterol alone or with dietary fat. Saturated fat is more atherogenic than unsaturated fat (3) and the severity of lesions may be enhanced by adding free fatty acids to the diet (4). In cholesterol-fed rabbits lesions can be observed in a relatively short time, usually within 2-3 months. Scebat (5) has shown long-term feeding of low levels of sterol can be almost as effective as feeding of high cholesterol diets (Table 1).

Atherosclerosis can also be induced in rabbits by feeding a semipurified diet containing 20-25% protein, 40-50% carbohydrate and 8-20% saturated fat (6-8). This diet must be fed for longer periods than the cholesterol containing diet, usually 6-10 months. Atherosclerosis is induced in chickens by feeding cholesterol plus fat. Insofar as the type of fat used is concerned, there is no universally clear cut effect of saturated fat. Some investigators find unsaturated fats to be more atherogenic than

Table 1. Response of Rabbits to Graded Dosages of Cholesterol*

	GROUP		
	I	II	III
Dose of cholesterol, mg/day	250	500	1000
Serum cholesterol, mg/dl			
Day 30	271	651	813
Day 90	250	1152	1072
Day 130	545	1103	1095
Average atherosclerosis			
Day 30	0.9	1.3	1.8
Day 90	-	2.3	3.0
Day 130	3.0	3.3	3.8

*After Scebat et al. (5).

Table 2. Influence of Peanut Oil and Related Oils on Cholesterol-
Induced Atherosclerosis in Rabbits*

Regimen⧧	No.	Serum Cholesterol (mg/dl)	Atheromata	
			Arch	Thoracic
Peanut oil	26	1339	1.94	1.23
PGF	29	1428	1.55	1.17
PGF/R	31	1723	1.40	0.94
Corn oil	30	1529	1.43	0.97
Peanut oil	27	1873	2.22	1.54
PNO/R	31	1833	1.31	1.05
Corn oil	28	1678	1.32	1.02

*After Kritchevsky et al. (28).
 All rabbits fed 2% cholesterol, 6% fat for 2 months. Aortas
 graded on 0-4 scale.
⧧PGF - Oil blended to resemble peanut oil minus 20:0 and 22:0.
 PGF/R - PGF interesterified with 20:0 and 22:0 glycerides to
 give composition identical to peanut oil.
 PNO/R - Autointeresterified (randomized) peanut oil.

saturated fats (9) and some find the reverse to be true (10).
Undernutrition renders chickens more susceptible to atherosclerosis
(11) as does a low protein diet (12). In the rat it is necessary
to take extreme physiological measures to induce lesions. Generally it is enough to ablate the thyroid either chemically or
surgically and to feed a diet rich in cholesterol and cholic acid
(13). If the dietary fat is saturated one can obtain thrombosis
(14, 15), if it is peanut oil, atherosclerosis is induced (14).
Hypophysectomy followed by a cholesterol-rich diet (16), saturated
fat fed to rats with essential fatty deficiency (17) or feeding
of high levels of Vitamin·D_2 and cholesterol (18) also lead to
atherosclerosis in the rat. Feeding cholesterol plus thiouracil
(19) or cholesterol plus a semipurified (20) will lead to atherosclerosis in the dog. The ease with which atherosclerosis can be
induced in primates is a function of the strain. Thus, cholesterol feeding is sufficient to cause atherosclerosis in the rhesus
(21) or squirrel (22) monkey but the cebus monkey, generally
requires cholesterol plus a diet low in sulfur amino acids (23) or
pyridoxine deficiency (24).

The effect of peanut oil in atherosclerosis is an aspect of
dietary induction of aortic lesions that merits discussion. Ahrens
(25) reported that peanut oil has an effect on serum cholesterol in
man which is similar to that of corn oil. Nonetheless this oil is
inordinately atherogenic for rhesus monkeys (26), rabbits (27) and
rats (14). Since peanut oil contains up to 6% of long chain
saturated fatty acids (arachidic, behenic, lignoceric) we examined
the possibility that a fat resembling peanut oil minus these fatty
acids would be less atherogenic and, indeed it was (27). Adding
arachidic and behenic acids to this fat (by interesterification)
increased its atherogenicity but not to the level seen with peanut
oil itself (28).

The interesterification process affects the atherogenicity of
peanut oil (28). Interesterification results in a fat in which
every component fatty acid is present in each position of the
triglyceride to approximately óne-third of its total concentration.
Interesterified, or randomized, peanut oil has the same iodine
value and fatty acid spectrum as native peanut oil but the structure of its component triglycerides differs from that of the
starting oil (29). The atherogenicity of peanut oil is as dependent on its structure as it is on its level of unsaturation. Our
findings are summarized in Table 2.

We have used a semipurified diet (40% sucrose, 25% casein,
14% coconut oil) to establish atherosclerosis in rabbits (8).
This diet permits us to examine the effects of variation of individual dietary components, thus glucose and lactose are less atherogenic than fructose or sucrose (30, 31); casein is more atherogenic than soy protein when the fiber is cellulose but the two

proteins are equivalent when the fiber is alfalfa (32); and coconut, butter and peanut oils are more atherogenic than corn oil (33). Using this semipurified we have examined the effects of trans fatty acids. Kummerow (34) has suggested that dietary trans fatty acids exert an untoward atherogenic effect. In experiments with swine Kummerow has shown trans fats to be atherogenic in one experiment (35) but not in another (36). McMillan and his colleagues fed rabbits cholesterol with oleic or elaidic acids (37), triolein or trielaidin (38) or cis-trans, trans-cis, or trans-trans linoleic acid (39). The trans fats were more cholesteremic but not more atherogenic. We fed low (3.2%) and high (6.0%) levels of elaidic to rabbits as part of the cholesterol-free, semipurified diet. The trans fat was more hypercholesteremic but no more atherogenic than the control diet. Furthermore it did not affect activity of hepatic glucose-6-phosphatase, fatty acid synthetase, malate dehydrogenase, monoamine oxidase or β-hydroxybutyrate dehydrogenase (40).

The effect of dietary protein deserves special mention. The earliest atherosclerosis experiments (1) were a test of protein effects but since then relatively few experiments have been devoted to this nutrient. Hamilton and Carroll (41) showed that animal proteins were more cholesteremic for rabbits than were vegetable proteins but there was a large variation within each group. We (32) have confirmed the earlier observation that soy protein is less atherogenic than casein (42). We have shown further (43, 44) that the addition of lysine to soy protein increases its hypercholesteremic and atherogenic properties. Beef protein is of the same order of atherogenicity as casein but its cholesteremic and atherogenic potential is significantly reduced when it is diluted 50% with textured vegetable protein (43).

The semipurified diet has also been fed to primates. In baboons fed diets in which only the carbohydrate was varied fructose was shown to be more sudanophilic than glucose (45). In vervet monkeys fructose was significantly more atherogenic than glucose (46). When 0.1% cholesterol was added to the semipurified diets it led to atherosclerosis, the lesions being more pronounced and more severe in the animals fed lactose (47).

We have studied a number of pharmaceutical agents as potential agents for lowering cholesterol and inhibiting atherosclerosis. Among the agents studied have been D-and L-thyroxine (48), W-1372 (49), probucol (50), colestipol (51), clofibrate (52), Linolexamide (53) and WY-14643 (54). Our data are summarized in Table 3. All the drugs were used in conjunction with a cholesterol-containing diet. The question of regression of atherosclerosis in primates has been discussed in detail by Dr. Wissler, whose paper appears elsewhere in these proceedings.

Table 3. Influence of Drugs on Experimental Atherosclerosis in Rabbits*

Regimen	Dose	No.	Serum Cholesterol (mg/dl)	Atheromata		Ref.
				Arch	Thoracic	
D-Thyroxine	0.5-1.0 mg/day	29	1374	1.78	1.34	48
L-Thyroxine	0.05 mg/day	25	1270	2.18	1.80	
Control	-	33	1913	2.76	1.90	
Clofibrate	0.3%	42	1851	1.90	1.29	52
Control	-	41	1996	2.29	1.38	
W-1372	2%	18	1178	1.19	0.67	49
Control	-	18	1166	1.81	1.17	
Probucol	1%	28	1283	1.57	0.93	50
Control	-	29	1889	2.05	1.38	
Colestipol	1%	30	1721	1.36	0.97	51
Control	-	29	1890	1.88	1.31	
Linolexamide	600 mg	22	1115	1.10	0.66	53
Control	-	19	1968	1.87	0.92	
WY-14643	5 mg/day	8	2029	1.50	0.90	54
Control	-	10	1575	2.30	1.30	

*All rabbits fed 2% cholesterol, 6% corn oil for 2 months; Aortas graded on 0-4 scale.

Atherosclerosis in rabbits does not seem to regress. McMillan et al. (55) fed rabbits 6 gm of cholesterol/week for 3 months. At the end of this period the average grade of atherosclerosis for 45 rabbits was 2.42. The rabbits were returned to commercial ration and several animals autopsied at regular intervals for 26 weeks. There were wide variations in levels of atheroma but at the 25th week the average atherosclerosis in 9 rabbits was 2.56. We have observed that when rabbits are returned to commercial ration after 8 weeks on an atherogenic regimen (2% cholesterol) atherosclerosis is exacerbated. When saturated fat is added to the post-cholesterol diet the lesions become even more severe; when unsaturated fat is added to the diet the exacerbation of lesions is reduced significantly but the lesions are more severe than those observed at the cessation of cholesterol feeding (56). Vles et al. made a similar observation (57). Agents such as choline (58), sitosterol (59), clofibrate (52) or thyroxine (48) have little or no effect on regression of established lesions. A striking regression of induced atherosclerosis in rabbits has been achieved by Vesselinovitch et al.(60) who exposed the rabbits to 100% oxygen for 2 hours each day. Return of the rabbits from an atherogenic to a stock diet increased atherosclerosis by 2% but stock diet plus oxygen decreased atherosclerosis by 23%. Cholestyramine reduced lesions by 43%, cholestyramine plus oxygen by 55%. Estradiol benzoate also reduced atherosclerosis by 43%, when given together with oxygen the reduction was 52%.

This presentation has covered briefly factors (diet, drugs) which affect induction and regression of experimental atherosclerosis. Among the dietary factors the influence of protein may be of prime importance. Interaction of dietary components may also have a marked effect on the course of atherogenesis.

ACKNOWLEDGEMENT

Supported, in part, by grants (HL-03299 and HL-05209) and a Research Career Award (HL-0734) from the National Institutes of Health.

REFERENCES

1. A. Ignatowski, Über die Wirkung des tierischen Eiweisses an Aorta und die parenchymatosen Organe der Kaninchen, Virchows Arch. Pathol. Anat. Physiol. Klin. Med. 198:248 (1909).
2. N. Anitschkow and S. Chalatow, Über experimentelle Cholesterensteatose und ihre Bedeutung für die Entstehung einiger pathologischer Prozesse, Z. Allg. Pathol. Pathol. Anat. 24:1 (1913).

3. D. Kritchevsky, A. W. Moyer, W. C. Tesar, J. B. Logan, R. A. Brown, M. C. Davies and H. R. Cox, Effect of cholesterol vehicle in experimental atherosclerosis, Am. J. Physiol. 178:30 (1954).

4. D. Kritchevsky, S. A. Tepper and J. Langan, Cholesterol vehicle in Experimental Atherosclerosis IV. Influence of heated fat and fatty acids, J. Atheroscler. Res. 2:115 (1962).

5. L. Scebat, J. Renais and J. Lenegre, Atherosclerose experimentale du lapin. Etude preliminaire, Rev. Atherosclerose 3:14 (1961).

6. G. F. Lambert, J. P. Miller, R. T. Olsen and D. V. Frost, Hypercholesteremia and atherosclerosis induced in rabbits by purified high fat rations devoid of cholesterol, Proc. Soc. Exp. Biol. Med. 97:544 (1958).

7. H. Malmros and G. Wigand, Atherosclerosis and deficiency of essential fatty acids, Lancet 2:749 (1959).

8. D. Kritchevsky and S. A. Tepper, Factors affecting atherosclerosis in rabbits fed cholesterol-free diets, Life Sci. 4:1467 (1965).

9. D. M. Tennent, M. E. Zonetti, H. Siegel, G. W. Kuron and W. H. Ott, The influence of selected vegetable fats on plasma lipid concentrations and aortic atheromatosis in cholesterol-fed and diethylstilbesterol-implanted cockerels, J. Nutr. 69:283 (1959).

10. J. Stamler, R. Pick and L. N. Katz, Saturated and unsaturated fats. Effects on cholesterolemia and atherogenesis in chicks fed high-cholesterol diets, Circ. Res. 7:398 (1959).

11. S. Rodbard, R. Pick and L. N. Katz, The rate of regression of hypercholesteremia and atherosclerosis in chicks. Effects of diet, pancreatectomy, estrogens and thyroid, Circ. 10:597 (1954).

12. J. Stamler, R. Pick and L. N. Katz, Effects of dietary protein and carbohydrate level on cholesterolemia and atherogenesis in cockerels on a high-fat, high-cholesterol mash, Circ. Res. 6:447 (1958).

13. W. S. Hartroft, J. H. Ridout, E. A. Sellars and C. H. Best, Atheromatous changes in aorta, carotid and coronary arteries of choline-deficient rats, Proc. Soc. Exp. Biol. Med. 81:384 (1952).

14. G. A. Gresham and A. N. Howard, The independent production of atherosclerosis and thrombosis in the rat, Brit. J. Exp. Path. 41:395 (1960).

15. R. F. Scott, E. S. Morrison, W. A. Thomas, R. Jones and S. C. Nam, Short term feeding of unsaturated versus saturated fat in the production of atherosclerosis in the rat, Exp. Molec. Pathol. 3:421 (1964).

16. P. R. Patek, S. Bernick, B. H. Ershoff and A. Wells, Induction of atherosclerosis by cholesterol-feeding in the rat, hypophysectomized rat, Am. J. Pathol. 42:137 (1963).

17. R. J. Morin, S. Bernick and R. B. Alfin-Slater, Effects of
    essential fatty acid deficiency and supplementation
    on atheroma formation and regression, J. Atheroscler. Res.
    4:387 (1964).
18. D. Aubert, J. C. Ferrand, B. Lacaze, O. Pepin, E. Panak and
    M. Podesta, Atherogenese experimentale chez le rat Wistar,
    Atheroscler. 20:263 (1974).
19. A. Steiner and F. E. Kendall, Atherosclerosis and arterio-
    sclerosis in dogs following ingestion of cholesterol and
    thiouracil, Arch. Pathol. 42:433 (1946).
20. H. Malmros and N. H. Sternby, Induction of atheroclerosis
    in dogs by a thiouracil-free semisynthetic diet contain-
    ing cholesterol and hydrogenated coconut oil, Progr.
    Biochem. Pharmacol. 4:482 (1968).
21. G. V. Mann and S. B. Andrus, Xanthomatosis and atherosclerosis
    produced by diet in an adult rhesus monkey, J. Lab. Clin.
    Med. 48:533 (1956).
22. M. R. Malinow, C. A. Maruffo and A. M. Perley, Experimental
    atherosclerosis in squirrel monkeys (saimiri sciurea),
    J. Path. Bact. 92:491 (1966).
23. G. V. Mann, S. B. Andrus, A. McNally and F. J. Stare, Experi-
    mental atherosclerosis in cebus monkeys, J. Exp. Med.
    98:195 (1953).
24. J. F. Rinehart and L. D. Greenberg, Pathogenesis of experi-
    mental arteriosclerosis in pyridoxine deficiency; with
    notes on similarities to human arteriosclerosis, Arch.
    Pathol. 51:12 (1951).
25. E. H. Ahrens, Jr., Nutritional factors and serum lipid levels,
    Am. J. Med. 23:928 (1957).
26. D. Vesselinovitch, G. S. Getz, R. H. Hughes and R. W. Wissler
    Atherosclerosis in the rhesus monkey fed three food fats,
    Atheroscler. 20:303 (1974).
27. D. Kritchevsky, S. A. Tepper, D. Vesselinovitch and R. W.
    Wissler, Cholesterol vehicle in experimental atherosclero-
    sis. II. Peanut Oil, Atheroscler. 14:53 (1971).
28. D. Kritchevsky, S. A. Tepper, D. Vesselinovitch and R. W.
    Wissler, Cholesterol vehicle in experimental atherosclero-
    sis. 13. Randomized peanut oil, Atheroscler. 17:225 (1973).
29. J. J. Myher, L. Marai, A. Kuksis and D. Kritchevsky, Acylgly-
    cerol structure of peanut oils of different atherogenic
    potential, Lipids 12:775 (1977).
30. D. Kritchevsky, P. Sallata and S. A. Tepper, Experimental
    atherosclerosis in rabbits fed cholesterol-free diets.
    2. Influence of various carbohydrates, J. Atheroscler.
    Res. 8:697 (1968).
31. D. Kritchevsky, S. A. Tepper and M. Kitagawa, Experimental
    atherosclerosis in rabbits fed cholesterol-free diets.
    3. Comparison of fructose and lactose with other
    carbohydrates, Nutr. Rep. Int. 7:193 (1973).

32. D. Kritchevsky, S. A. Tepper, D. E. Williams and J. A. Story, Experimental atherosclerosis in rabbits fed cholesterol-free diets. 7. Interaction of animal or vegetable protein with fiber, Atheroscler. 26:397 (1977).

33. D. Kritchevsky, S. A. Tepper, H. K. Kim, J. A. Story, D. Vesselinovitch and R. W. Wissler, Experimental atherosclerosis in rabbits fed cholesterol-free diets. 5. Comparison of peanut, corn butter and coconut oils, Exp. Molec. Pathol. 24:375 (1976).

34. F. A. Kummerow, Current studies on relation of fat to health, J. Am. Oil Chem. Soc. 51:255 (1974).

35. F. A. Kummerow, Lipids in atherosclerosis, J. Food Sci. 40:12 (1975).

36. R. L. Jackson, J. D. Morrissett, H. J. Pownall, A. M. Gotto, Jr., A. Kamio, H. Imai, R. Tracy and F. A. Kummerow, Influence of dietary trans fatty acids on swine lipoprotein composition and structure, J. Lipid Res. 18:183 (1977).

37. B. I. Weigensberg, G. C. McMillan and A. C. Ritchie, Elaidic acid: Effect on experimental atherosclerosis, Arch. Pathol. 72:358 (1961).

38. G. C. McMillan, M. D. Silver and B. I. Weigensberg, Elaidinized olive oil and cholesterol atherosclerosis, Arch. Pathol. 76:106 (1963).

39. B. I. Weigensberg and G. C. McMillan, Serum and aortic lipids in rabbits fed cholesterol and linoleic acid steroisomers, J. Nutr. 83:314 (1964).

40. H. Ruttenberg, N. A. Little, L. M. Davidson and D. Kritchevsky, Influence of dietary trans fatty acid on atherosclerosis in rabbits, Fed. Proc. 39:1039 (1980).

41. K. K. Carroll and R. M. G. Hamilton, Effects of dietary protein and carbohydrate on plasma cholesterol levels in relation to atherosclerosis, J. Food Sci. 40:18 (1975).

42. D. R. Meeker and H. D. Kesten, Effect of high protein diets on experimental atherosclerosis of rabbits, Arch. Pathol. 31:147 (1941).

43. D. Kritchevsky, Vegetable protein and atherosclerosis, J. Am. Oil Chem. Soc. 56:135 (1979).

44. S. K. Czarnecki and D. Kritchevsky, The effect of dietary protein on lipoprotein metabolism and atherosclerosis in rabbits, J. Am. Oil Chem. Soc. 56:388A (1979).

45. D. Kritchevsky, L. M. Davidson, I. L. Shapiro, H. K. Kim, M. Kitagawa, S. Malhotra, P. P. Nair, T. B. Clarkson, I. Bersohn, P. A. D. Winter, Lipid Metabolism and experimental atherosclerosis in baboons: Influence of cholesterol-free, semi-synthetic diets, Am. J. Clin. Nutr. 27:29 (1974).

46. D. Kritchevsky, L. M. Davidson, H. K. Kim, D. A. Krendel, S. Malhotra, J. J. van der Watt, J. P. du Plessis, P. A. D. Winter, T. Ipp, D. Mendelsohn and I. Bersohn,

Influence of semipurified diets on atherosclerosis in
African green monkeys, Exp. Mol. Pathol. 26:28 (1977).

47. D. Kritchevsky, L. M. Davidson, H. K. Kim, D. A. Krendel,
S. Malhotra, D. Mendelsohn, J. J. van der Watt, J. P.
du Plessis and P. A. D. Winter, Influence of type of
carbohydrate on atherosclerosis in baboons fed semipurified diets plus 0.1% cholesterol, Am. J. Clin. Nutr.
33:1869 (1980).

48. D. Kritchevsky, J. L. Moynihan, J. Langan, S. A. Tepper and
M. L. Sachs, Effects of D-and L-Thyroxine and of D-and
L-3, 5, 3'-triiodothyronine on development and regression
of experimental atherosclerosis in rabbits, J. Atheroscler.
Res. 1:211 (1961).

49. D. Kritchevsky, P. Sallata and S. A. Tepper, Effect of N-α-
phenylpropyl-N-benzyloxy Acetamide (W1372) on experimental atherosclerosis in rabbits, Proc. Soc. Exp. Biol. Med.
132:303 (1969).

50. D. Kritchevsky, H. K. Kim and S. A. Tepper, Influence of
4,4'-(isopropylidenedithio)bis(2,6-di-t-butylphenol)
(DH581) on experimental atherosclerosis in rabbits,
Proc. Soc. Exp. Biol. Med. 136:1216 (1971).

51. D. Kritchevsky, H. K. Kim and S. A. Tepper, Effect of Colestipol (U-26597A) on experimental atherosclerosis in
rabbits, Proc. Soc. Exp. Biol. Med. 142:185 (1973).

52. D. Kritchevsky, P. Sallata and S. A. Tepper, Influence of
ethyl p-chlorophenoxyisobutyrate (CPIB) upon establishment
and progression of experimental atherosclerosis in rabbits.
J. Atheroscler. Res. 8:755 (1968).

53. D. Kritchevsky and S. A. Tepper, Linolexamide (N-cyclohexyl-
linoleamide) in experimental atherosclerosis in rabbits,
J. Atheroscler. Res. 7:527 (1967).

54. D. Kritchevsky, D. E. Moses and S. A. Tepper, Influence of
(4-chloro-(2,3-xylidino)-2-pyrimidinylthio) Acetic Acid
(WY 14643) on experimental atherosclerosis in rabbits,
Artery 1:10 (1974).

55. G. C. McMillan, L. Horlick and G. L. Duff, Cholesterol content of aorta in relation to severity of atherosclerosis,
Arch. Pathol. 59:285 (1955).

56. D. Kritchevsky and S. A. Tepper, Cholesterol vehicle in
experimental atherosclerosis. V. Influence of fats and
fatty acids on pre-established atheromata, J. Atheroscler.
Res. 2:471 (1962).

57. R. O. Vles, J. Buller, J. J. Gottenbos and H. J. Thomasson,
Influence of type of dietary fat on cholesterol-induced
atherosclerosis in the rabbit, J. Atheroscler. Res. 4:170
(1964).

58. G. L. Duff and G. F. Meissner, Effect of choline on the
development and regression of cholesterol atherosclerosis
in rabbits, Arch. Pathol. 57:329 (1954).

59. W. T. Beher, W. L. Anthony and G. D. Baker, Effects of beta-sitosterol on regression of cholesterol atherosclerosis in rabbits, Circ. Res. 4:485 (1956).
60. D. Vesselinovitch, R. W. Wissler, K. Fisher-Dzoga, R. Hughes and L. Dubien, Regression of atherosclerosis in rabbits. 1. Treatment with low-fat diet, hyperoxia and hypolipidemic agents, Atheroscler. 19:259 (1974).

ON THE SELECTION OF ANIMAL SPECIES AS MODELS FOR ATHEROSCLEROSIS

RESEARCH, WITH PARTICULAR REFERENCE TO THE COMMON MARMOSET

(CALLITHRIX JACCHUS)

M. John Chapman, Patricia Forgez, Sonia Goldstein and
Fergus McTaggart*
Unit 35,Groupe de Recherches sur le Métabolisme des Lipides,
Institut National de la Santé et de la Recherche Médicale,
Hôpital Henri Mondor, 94010 CRETEIL, France,   and
*Pharmaceutical Division,Imperial Chemical Industries Ltd.,
Alderley Park, MACCLESFIELD, Cheshire SK10 4TG,   U.K.

Much has been written in recent years of the complex and multiple processes implicated in the development of atherosclerosis.
Despite such intense activity in this field however, it is evident
from the present Conference that controversy still sorrounds many
aspects of this insidious disease. One may especially cite questions
sorrounding the nature and relative importance of the various factors
which may contribute to the initiation of the atherogenic process
itself, as well as those relevant to its regression.

As a consequence of several extensive epidemiological surveys
in human populations, such as that performed over a period of years
in Framingham,Massachusetts,(Kannel et al.,1971), we have come to
recognise the importance of elevated serum lipid (and particularly
cholesterol) levels as major risk factors for vascular disease.
The intractability and undesirability of humans as subjects for
detailed study of the disease process has tended however to restrict
researchers to use of post-mortem tissue. Such limitations thus
prompted the search for alternative species as experimental models
for study of the dynamic facets of lesion development.

The earliest use of an animal species for atherosclerosis
research was perhaps that of the rabbit by Anitschkow in the 1920's

(Anitschkow,1967). Indeed his studies in rabbits fed cholesterol-
supplemented diets provided a major stimulus for subsequent investigations of the pathogenesis of the disease in a wide variety of
animals rendered hypercholesterolemic by dietary means.

To date, members of almost all the major classes of vertebrates
have been the subject of atherosclerosis-related study. Among them,
representations of the fishes (salmon, Oncorhynchus sp.; rainbow
trout, Salmo gairdnerii), birds (pigeon, Columba sp.; chicken, Gallus
domesticus; turkey, Meleagris galapavo) and mammals (rat, Rattus
norvegicus; rabbit, Oryctolagus cuniculus; dog, Canis familiaris;
pig, Sus domesticus) have been the subject of most attention. To
these species may be added the non-human primates (especially Old
and New World monkeys), which as a group, account for possibly
the largest number of experimental studies within the past decade;
many of these have emanated from the Primate Research Centers in the
United States (for reviews, see Clarkson et al.,1976, Wissler,1979
and Malinow,1980).

Unlike Western man, in which atherosclerotic disease is apparently indigenous, the vast majority of animal species only rarely
appear to present with spontaneous lesions in their natural habitat.
This has led to the inbreeding of some species, such as certain
strains of quail (Chapman et al.,1976), pigeons (Clarkson et al.,
1976), mice (Roberts and Thompson,1976) and monkeys (Clarkson et al.,
1976), in order to derive animals with increased susceptibility to
experimental atherosclerosis. Nonetheless, the induction of arterial
lesions in such inbred strains still suffers from the same drawback
as their random-bred counterparts,i.e. relatively drastic measures,
usually involving administration of massive amounts of cholesterol
and saturated fat, must be employed in order to induce arterial
lesion formation within a comparatively short time period.

If we accept the thesis that diet-induced hypercholesterolemia
has been the most common means of producing experimental atherosclerosis in animals, then it would appear reasonable to assume
that some fundamental relationship exists between the nature of the
circulating cholesterol-transporting molecules, in this case the
serum lipoproteins, and the nature and anatomical distribution of
the arterial lesions. Furthermore, since the hypercholesterolemic
animal species which have been studied have been assumed to represent
valid models for the evaluation of the disease process as it sup-
posedly occurs in man, then it has to be further assumed that their

lipoprotein and apolipoprotein profiles correspond in large part to
that typical of those markedly hypercholesterolemic subjects showing
the greatest susceptibility to atherosclerosis,i.e. those with elevated LDL (low-density lipoprotein) levels and classified as Type II by
Fredrickson et al.(1978).

Examination of data in the literature will however reveal this
not to be the case. Thus, relatively few studies of experimental
atherosclerosis have been performed on animals which normally display significant LDL levels (> 100 mg/dl) and which respond to
dietary fat and cholesterol by markedly and almost exclusively elevat-
ing the levels of this lipoprotein class. In fact, the great majority
of species employed for such studies to date may be termed "HDL (high-
density lipoprotein)-animals", in the sense that under normal dietary
conditions, a major portion of the total circulating cholesterol
(50% or more in certain species, as in the rat) is transported in the
HDL rather than in the LDL class as in man; the pig is one of the
few exceptions (for a review, see Chapman,1980). When such animals
(e.g. rat, rabbit, dog and pig) are fed high fat, high-cholesterol
diets, they typically respond by producing large amounts of "abnormal"
lipoproteins, amongst which B-migrating VLDL (very-low density lipoprotein) and a cholesterol-rich form of HDL containing the arginine-
rich apoprotein and termed HDL$_c$, are prominent (for reviews, see
Mahley,1978, and Shore and Shore,1976). Thus the lipoprotein pattern
in these cholesterol-fed animals is often more akin to the relatively
rare human disease of Type III hyperlipoproteinemia rather than to
the predominant Type II condition. Moreover, these comments extend
to certain of the nonhuman primates, as exemplified by the studies
of Mahley et al.(1976) in hypercholesterolemic Erythrocebus patas
monkeys. A further anomaly which is typically observed in some
cholesterolemic Old World monkeys (Clarkson et al.,1976), and notably
the rhesus (Mataca mulatta) and cynomolgous (Macaca fascicularis)
monkeys concerns the large amounts of highly heterogeneous LDL, a
characteristic quite distinct from that of the LDL distribution
generally seen in Type II subjects.

These observations prompted us to search for alternative species
in which the hypercholesterolemic lipoprotein and apolipoprotein
profiles might approach those of Type II individuals more closely.

An early candidate for our attention was the guinea pig (Cavia
porcellus), which although a rodent and herbivore, exhibited low-

density, apolipoprotein B-containing particles as its principal serum
lipoprotein class when fed a chow diet; concentrations of HDL were
low (< 20mg/dl) (Chapman,1980). This animal typically responds to
acute administration (~ 1 week) of a diet supplemented with cholesterol (1.6%) and unsaturated fat (16%) by increasing the levels
of its LDL almost exclusively (Chapman and Mills,1977; Mills et al.,
1972). Several aspects of the metabolism and structure of the LDL
in such hypercholesterolemic animals closely resemble those occurring
in certain cases of familial hypercholesterolemia in man (Mills et
al.,1972; Mills and McTaggart,1974; Mills et al.,1976). These phenomena are not to be confused with those which characterise longer
periods (up to 3 months) of cholesterol feeding, when several species
of abnormal lipoprotein appear in conjunction with a hemolytic anemia
(Sardet et al.,1972). Indeed, such effects have limited the use of
Cavia as an animal model for studies of the development of experimental atherosclerosis at the cellular level.

More recently, a New World primate, the common Marmoset (Callithrix jacchus), has attracted our interest. This species, like the
spider and squirrel monkeys (Ateles geoffroyi and Saimiri sciureus
respectively), displays LDL levels which approach those of man a
good deal more closely than those of several Old World varieties
(Chapman et al.,1979). We considered this aspect alone sufficient to
warrant further studies of the lipoprotein and apolipoprotein profile
in normolipidemic marmosets: such investigations were the subject of
a recent report from our laboratory (Chapman et al.,1979), and served
to reveal that many qualitative and quantitative aspects of the profile resembled those typical of man. It was however notable that,
like several other Old and New World monkeys but with the exception
of the spider monkey (Srinivasan,1976), the marmoset transports
some 50% of its total cholesterol in the form of HDL. Moreover, the
predominant form of HDL present is HDL_2, and not HDL_3 as in man
(HDL_2 : HDL_3 ~ 3.5:1 in marmoset ~ 1:3 in man). Such elevated HDL_2
concentrations, common also to the rhesus and squirrel monkeys
(Scanu et al.,1973; Illingworth,1975), may however make these species
good models for investigation of the metabolism of this lipoprotein
class,and more especially, of that subclass of HDL_2 judged highly
protective against atherosclerosis in man (Anderson et al.,1978).

The occurrence of spontaneous atherosclerosis in wild marmosets
is apparently rare, making this a moderately resistant species (Dreizen et al.,1973); the disease occurs more frequently in wild spider

and squirrel monkeys. Nonetheless, the cotton-top marmoset (Saguinus oedipus) may be rendered hypercholesterolemic and atherosclerosis-susceptible by the expedient of feeding a 5% cholesterol, 23% lard and 2% corn oil-containing diet (Dreizen et al.,1973). Intimal fat deposition was not observed until the animals had received this diet for at least a year; at this time, lesions in the lingual arteries and distal extensions of the coronary arteries were prominent and antedated those in the aorta.

As a preface to investigation of the lipid transport system in hypercholesterolemic marmosets and of the mechanisms operative in their experimental atherosclerosis, we are presently characterising the vectors of lipid transport, i.e. the serum apolipoproteins, in this species. To date, we have demonstrated that the major apolipoproteins of normolipidemic animals are analogous to the human B and AI proteins. Minor differences in the structure of these apoproteins between the two species occur; thus, marmoset LDL (> 95% of the protein moiety of marmoset LDL of d. 1.027-1.055 g/ml is apo-B) is only partially immunochemically identical with human LDL, while marmoset apo-AI displays a range of pI values upon analytical isoelectric focussing which are distinct from those of its human counterpart (Figure 1). Thus, marmoset apo-AI is a slightly more acidic protein.

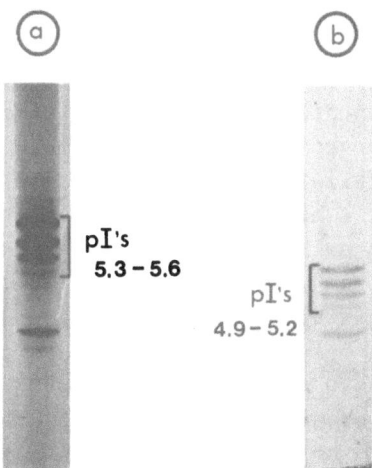

Fig.1 - Isoelectric focussing patterns in polyacrylamide gel of the urea-soluble apolipoproteins of HDL in (a) man and (b) the marmoset. Values for the pI of each band were determined by reference to a plot of pH against gel length made on control gels lacking protein.

In conclusion, emphasis in this discussion has been given to
the selection of animal models for use in atherosclerosis research,
and it has been suggested that the 'basal' lipoprotein and apolipoprotein profile in the animal species and that of normolipidemic man
should show an overall likeless. Moreover, if further understanding
of the cellular pathobiology underlying lesion formation and cholesterol deposition in hypercholesterolemic (Type II) individuals is
to be gained, then we judge it desirable that the animal responds
to dietary cholesterol so as to preferentially elevate its circulating LDL levels. These comments appear warranted to us, since to dace,
the great majority of investigations of the role of lipoproteins
in experimental atherosclerosis have been performed in species whose
hypercholesterolemic lipoprotein distribution is more reminiscent
of the rare Type II disease rather than that of the more prevalent
Type II condition. Finally, it must be said that a highly useful
animal model of atherosclerosis would incorporate both elevated
LDL levels and a defect in the functioning of its cellular LDL
receptors (Goldstein and Brown,1977). Ultimately, a complete understanding of atherogenesis will, in all likelihood, be derived from
summation of studies in many different species, each contributing
by virtue of a specific perturbation(s) at the cellular or lipoprotein level, or both. Eventually therefore, a rational approach
to prevention of the disease may be possible.

References

Anderson,D.W.,Nichols,A.V.,Pan,S.S.,and Lindgren,F.T., 1978 , High
    density lipoprotein distribution: resolution and determina-
    tion of three major components in a normal population sample,
    Atherosclerosis, 29:161

Anitschkow,N.N., 1967 , A history of experimentation on arterial
    atherosclerosis in animals, in "Cowdry's Atherosclerosis : a
    survey of the problem",H.T. Blumenthal, ed.,C.C. Thomas,
    Springfield,Illinois,U.S.A.

Chapman,M.J., 1980, Animal lipoproteins: chemistry, structure and
    comparative aspects, J.Lipid Res., 21:789

Chapman,M.J.,McTaggart,F.,and Goldstein,S., 1979, Density distribu-
    tion, characterisation and comparative aspects of the major
    serum lipoproteins in the Common Marmoset (Callithrix jacchus),
    a New World primate with potential use in lipoprotein research,
    Biochemistry , 18:5096

Chapman,M.J.,and Mills,G.L., 1977, Characterisation of the serum
    lipoproteins and their apoproteins in hypercholesterolemic
    guinea pigs, Biochem.J., 167:9
Chapman,K.P.,Stafford,W.W.,and Day,C.E., 1976, Animal models for
    experimental atherosclerosis produced by selective breeding
    of Japanese Quail, in "Atherosclerosis Drug Discovery", C.E.
    Day,ed., Plenum Press,New York
Clarkson,T.B.,Prichard,R.W.,Bullock,B.C.,St.Clair,R.W.,Lehner,N.D.
    M.,Jones,D.C.,Wagner,W.D., and Rudel,L.L., 1976, Pathogenesis
    of atheroslerosis; some advances from using animal models,
    Exp.Mol.Pathol., 24:264
Dreizen,S.,Levy,B.M.,and Bernick,S., 1973, Diet-induced atheroscle-
    rosis in the marmoset, Proc.Soc.Exp.Biol.Med., 143:1218
Fredrickson,D.S.,Goldstein,J.L.,and Brown,M.S., 1978, The familial
    hyperlipoproteinemias, in "The Metabolic Basis if Inherited
    Disease", J.B. Steinbury,J.B. Wyngaarden and D.S. Fredrickson,
    eds.,4th edition,McGraw-Hill, New York
Goldstein,J.L.,and Brown,M.S., 1977, The low density lipoprotein
    pathway and its relation to atherosclerosis, Ann.Rev.Biochem.,
    46:897
Illingworth,D.R., 1975, Metabolism of lipoproteins in nonhuman
    primates. Studies on the origin of low density lipoprotein
    apoprotein in the plasma of the squirrel monkey, Biochim.
    Biophys.Acta, 338:38
Kannel,W.B.,Castelli,W.P.,Gordon,T.,and McNamara,P.M., 1971, Serum
    cholesterol, lipoproteins and the risk of coronary heart disease,
    Ann.Intern.Med., 74:1
Mahley,R.W., 1978, Alterations in plasma lipoproteins induced by
    cholesterol feeding in animals including man, in "Disturbances
    in Lipid and Lipoprotein Metabolism", J.M. Dietschy,A.M. Gotto
    and J.A. Ontko,eds.,Am.Physiol.Soc., Bethesda
Mahley,R.W.,Weisgraber,K.H.,and Innerarity,T.L., 1976, Atherogenic
    hyperlipoproteinemia induced by cholesterol feeding in the
    patas monkey, Biochemistry, 15:2979
Malinow,M.R., 1980, Atherosclerosis- Regression in Nonhuman primates,
    Circ.Res., 46:311
Mills,G.L.,Chapman,M.J.,and McTaggart,F., 1972, Some effects of diet
    on guinea pig serum lipoproteins, Biochim.Biophys.Acta, 260:401
Mills,G.L.,Taylaur,C.E.,and Chapman,M.J., 1976, Low-density lipo-
    proteins in patients homozygous for familial hyperbetalipo-
    proteinemia, Clin.Sci.Mol.Med., 51:221

Mills,G.L.,and McTaggart,F., 1974, Characterization of experimental
    Type II hyperlipoproteinemia in guinea pigs, in "Atherosclerosis
    III", G. Schettler and A. Weizel,eds.,Springer-Verlag,Berlin
Roberts,A.,and Thompson,J.S., 1976, Inbred mice and their hybrids
    as an animal model for atherosclerosis research, in "Athero-
    sclerosis Drug Discovery",C.E. Day,ed., Plenum Press,New York
Sardet,C.,Hansma,H.,and Ostwald,R., 1972, Characterization of guinea
    pig plasma lipoproteins: the appearance of new lipoproteins
    in response to dietary cholesterol, J.Lipid Res., 13:624
Scanu,A.M.,Edelstein,C.,Vitello,L.,Jones,R.,and Wissler,R.W., 1973,
    The serum high density lipoproteins of Macacus rhesus I. Isola-
    tion, composition and properties, J.Biol.Chem., 218:7648
Shore,B.,and Shore,V., 1976, Rabbits as a model for the study of hyper-
    lipoproteinemia and Atherosclerosis, in "Atherosclerosis Drug
    Discovery", C.E. Day,ed.,Plenum Press, New York.
Srinivasan,S.R.,Smith,C.C.,Rashakrishnamurthy,B.,Wolf,R.H.,and Beren-
    son,G.S., 1976, Phylogenetic variability of serum lipids and
    lipoproteins in nonhuman primates fed diets with different con-
    tents of dietary cholesterol, Adv.Exp.Biol.Med., 67:65
Wissler,R.W., 1979, Evidence for regression of advanced atheroscle-
    rotic plaques, Artery, 5:398

SMOOTH MUSCLE METABOLIC REACTIVITY IN ATHEROGENESIS: LDL METABOLISM

AND RESPONSE TO SERUM MITOGENS DIFFER ACCORDING TO PHENOTYPE

Julie H. Chamley-Campbell[a], Paul Nestel[a] and
Gordon R. Campbell[b]

[a]Baker Medical Research Institute, Commercial Road,
Prahran, Victoria, 3181. Australia
[b]Department of Anatomy, University of Melbourne,
Parkville, Victoria, 3052. Australia

INTRODUCTION

Smooth muscle is the only cell type present in the media of mammalian arteries[1], and is therefore responsible for maintaining tension via contraction-relaxation and arterial integrity by proliferation and synthesis of connective tissue elements[2]. To accomplish this multiplicity of functions the smooth muscle cell is capable of expressing a range of phenotypes[3]. At one end of the spectrum is the smooth muscle cell whose major function is contrac - tion (contractile phenotype) with its cytoplasm filled with thick and thin myofilaments[4,5,6]. The vast majority of smooth muscle cells in the media of an adult aorta are in this phenotype. At the other end of the spectrum is the smooth muscle cell whose cytoplasm lacks thick filaments but contains scattered bundles of thin filaments and large amounts of rough endoplasmic reticulum, free ribosomes and Golgi, which are organelles associated with synthesis of both extracellular and intracellular material[3]. These cells are in a synthetic state (hence the term synthetic phenotype) and are present in developing and regenerating smooth muscle tissues[7], and also in atherosclerotic plaques[8,9].

In this paper we will show in primary cultures of enzyme dis- persed aortic smooth muscle that i) smooth muscle cells can undergo a reversible change of phenotype; ii) smooth muscle in the synthetic state will divide when challenged with whole blood serum whereas contractile state cells are unresponsive; iii) the ability of smooth muscle to degrade low density lipoprotein (LDL) differs according to the phenotypic state of the cells.

115

CHANGE IN PHENOTYPE

The thoracic aortic media of 1-5 year old slaughterhouse swine
(J.H. Ralph and Sons, Melbourne) is completely dispersed into
single smooth muscle cells by collagenase and elastase and seeded
into primary culture at known cell concentrations[5,10]. Approximately 60 million viable cells are obtained from one aorta. In
the first six days of culture the cells are in the contractile
state, that is, they respond to mechanical or electrical stimulation or the addition of angiotensin II at 10^{-7}g/ml by a slow
contraction. Ultrastructurally they closely resemble smooth
muscle cells in the intact pig aortic media with their cytoplasm
filled with bundles of thick and thin myofilaments[3,4,6]. They
stain intensely with FITC-labelled antibodies to the heavy chain
of smooth muscle myosin[11,12]. This antibody does not stain endothelial cells or fibroblasts and therefore is useful in checking
the cellular purity and homogeneity of medial cultures[12,13].

After 6 to 8 days in culture, the isolated smooth muscle cells
seeded at less than 10^6cells/ml undergo a spontaneous change in
phenotype to the synthetic state. Thick, myosin-containing
filaments can no longer be demonstrated and the cytoplasm is filled
with organelles involved with synthesis.[3,4,5,6] They do not
contract in response to stimulation, and lose their staining
reaction with the smooth muscle myosin antibody concomitant with
the loss of thick filaments[3,4,5,11].

RESPONSE TO SERUM MITOGENS

In the presence of 5% whole blood serum, before day 7 in
culture, only 1-3% of cells take up ^{3}H-thymidine into their nuclei
during a 4 hour pulse period.[10] However, once modulation to the
synthetic state has occurred the number of cells incorporating
[3]H-thymidine into DNA and thus preparing to divide increases to
10-20% (Fig. 1). This then increases linearly with time in culture
as progressively more cells are recruited into the cell cycle.
When the cells of sister cultures are counted in a haemocytometer
each day, the numbers remain constant until one to two days after
modulation of phenotype and increased incorporation of ^{3}H-thymidine.
The number of cells per dish then increases logarithmically until
confluence is achieved at which time cell proliferation ceases
(Fig. 2).

In the absence of the platelet-derived growth factor, that is
with 5% platelet deficient serum (PDS), the isolated smooth muscle
cells still modulate their phenotype on day 6 to 8, but less than
15% incorporate ^{3}H-thymidine into DNA and there is little increase
in cell number[10]. Similarly, in the total absence of serum, the
cells undergo a change in phenotype on day 6 to 8 but do not
proliferate. In the presence of 5% hyperlipemic whole blood serum

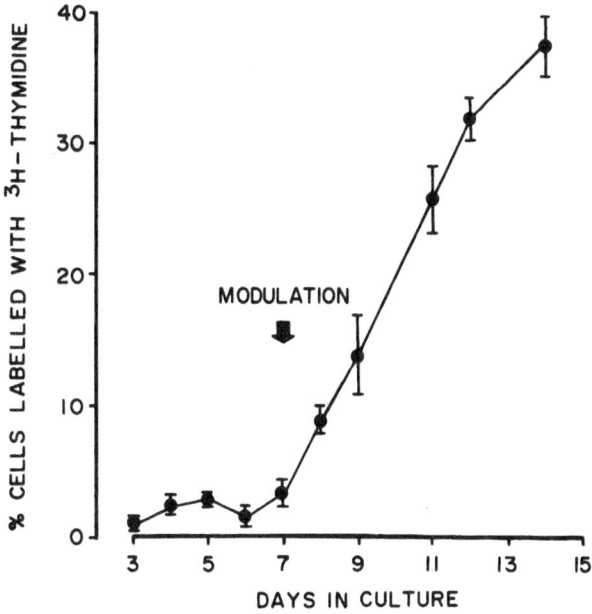

Fig. 1. Percentage of cells labelled with ^{3}H-thymidine during a 4 hour pulse period each day in culture in the presence of 5% whole blood serum. Each point represents the mean of 3 dishes ± SD.

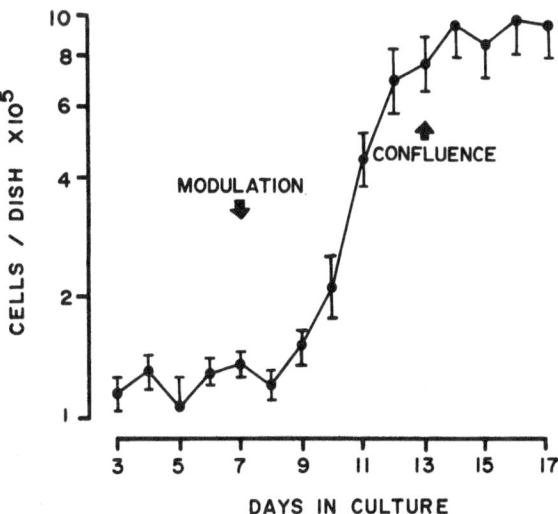

Fig. 2. Growth curve in the presence of 5% whole blood serum. Each point represents the mean of 3 dishes ± SD.

there is enhanced proliferation of synthetic state cells while
contractile state cells remain quiescent[10].

To summarize so far:

1.   Smooth muscle in the contractile phenotype (as it usually
exists in the adult blood vessel wall) cannot be stimulated to
divide by the platelet-derived growth factor, hyperlipemic LDL or
other serum derived factors.

2.   Phenotypic modulation from the contractile state to the
synthetic state is a necessary prerequisite for most smooth muscle
cells to become responsive to mitogens from serum.

3.   Neither platelet nor plasma derived factors are involved in
the modulation process.

If we can relate these <u>in vitro</u> results to the <u>in vivo</u>
situation, then knowledge of what controls smooth muscle phenotype
is important in understanding the pathogenesis of atherogenesis.

REVERSIBILITY OF PHENOTYPIC CHANGE

When enzyme isolated smooth muscle cells are plated at 10^6
cells/ml so that they form a confluent monolayer from day 1 in
culture, they do not undergo change in phenotype to the synthetic
state on day 7 but remain in the contractile state. If cells are
seeded at 5×10^4 to 1×10^5/ml, they modulate their phenotype on
day 7, proliferate until confluency is achieved on days 12-14, then
return to the contractile state (Fig. 2). However, if the cells
are seeded at 1×10^3 to 5×10^3/ml they modulate on day 7 and
proliferate but confluency is not achieved until about 3 weeks.
Under these conditions the cells do not return to the contractile
state but appear permanently in the synthetic state[14]. Cells which
have been subcultured several times also are permanently in a
synthetic state and are immediately responsive to serum mitogens
until rendered senescent through multiple cell division[15].

We suggest that these phenomena in culture may be analogous to
the atherosclerotic lesion, with the final phenotype of the plaque
smooth muscle cells dependent on the extent of the arterial insult
and thus the length of time these cells are in the synthetic phenotype undergoing proliferation. That is, after limited injury to
the arterial wall the cells modulate, migrate and proliferate,
rapidly restoring the integrity of the wall. The cells then return
to the contractile phenotype and must again modulate before being
capable of division. However, if the injury is large or repeated
so that the cells are in the synthetic state for a considerable
period undergoing proliferation, they do not return to the
contractile state and are therefore permanently responsive to serum

mitogens. Any further injury, no matter how small would then evoke an immediate proliferative response.

LDL METABOLISM IN CONTRACTILE VERSUS SYNTHETIC CELLS

Extensive lipid accumulation is a characteristic feature of atheromatous plaques and for this reason we were interested in comparing LDL metabolism of smooth muscle cells in the contractile versus synthetic phenotype. The methods used have been reported in detail [16],[17]; briefly, triplicate cultures are incubated for 20 hours in 5% lipoprotein deficient serum to maximize the number of LDL receptors, then incubated for 3 hours in 20 µg/ml I^{125} labelled LDL at 37°C. The amount (expressed as ng I^{125} LDL protein) bound to the high affinity receptors on the cell surface, internalized and degraded during this period is measured and expressed per 10^6 cells or per mg cell protein.

These parameters were measured in sparsely seeded pig aortic smooth muscle cell cultures undergoing phenotypic modulation on day 7 then proliferating to confluency on day 20 with no return to the contractile state. The binding of I^{125}-LDL decreased by a factor of two upon modulation to the synthetic state then remained low for the extent of the culture period while internalization remained constant throughout[17] (Fig. 3). The rate of degradation was high, but slowly decreasing with time in culture while the cells were in the contractile state, with a sudden and sharp decrease at the time of modulation to about one fourth the original level. The rate of degradation remained low while the cells were in the synthetic state even after the cells were confluent and had ceased proliferation.

The decrease in degradation is related to the modulation in phenotype rather than a response to culture since the rate of I^{125}-LDL degradation of cells seeded at 1 x 10^5/ml returns to approximately the original level after return to the contractile state upon confluency on day 13 (Fig. 4).

We conclude that smooth muscle cells in the synthetic phenotype (as many exist in atherosclerotic plaques) have a decreased ability to degrade certain lipoproteins than do those in the contractile state (as present in the media). This results in lipid accumulation which is evident morphologically when cells are grown in the presence of 5% hyperlipemic serum. Contractile state smooth muscle cells remain relatively free of lipid accumulation (Fig. 5) while synthetic state cells develop large lipid droplets (Fig. 6). This altered ability to metabolise lipoprotein by synthetic state smooth muscle may explain why some cells of the atherosclerotic lesion accumulate excessive amounts of lipid to become "foam cells".

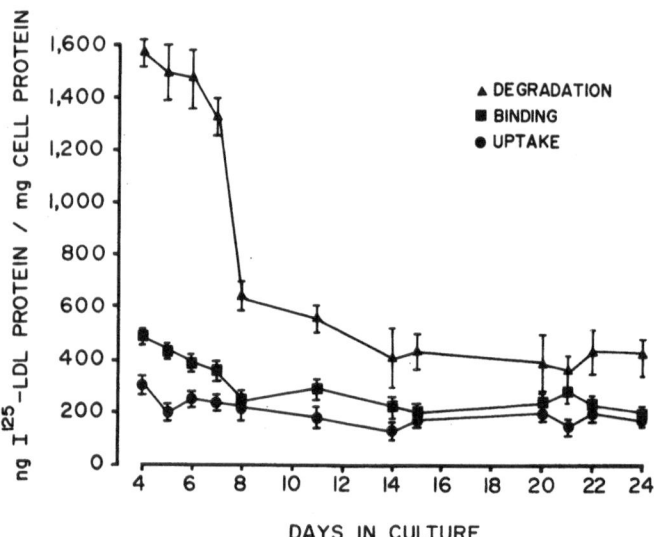

Fig. 3. Uptake,binding and degradation of I^{125} LDL with time in
culture. Modulation of smooth muscle phemotype occurred on day 7,
followed by proliferation then confluence on day 20 with no return
to the contractile state. Cell seeding 2.7 x 10^3/ml. Each point
represents the mean of 3 dishes $\pm$ SD.

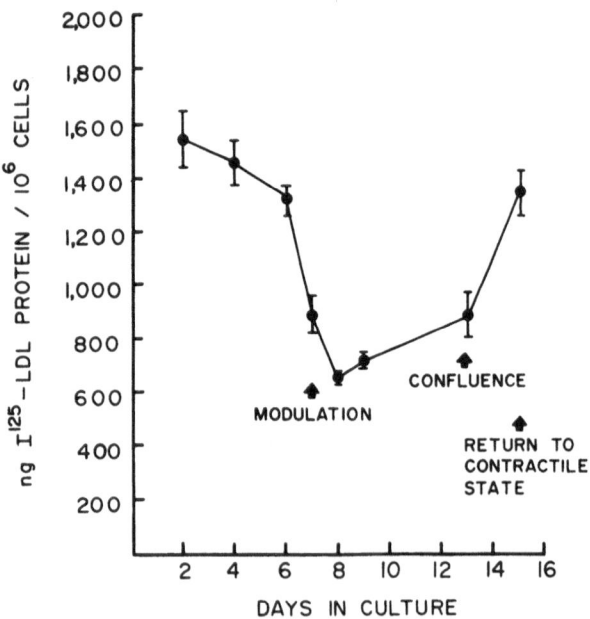

Fig. 4. Degradation of I^{125} LDL with time in culture. Confluence was
achieved on day 13 with return to the contractile state on day 15.
Cell seeding 1 x 10^5/ml. Each point represents the mean of 3 dishes
$\pm$ SD.

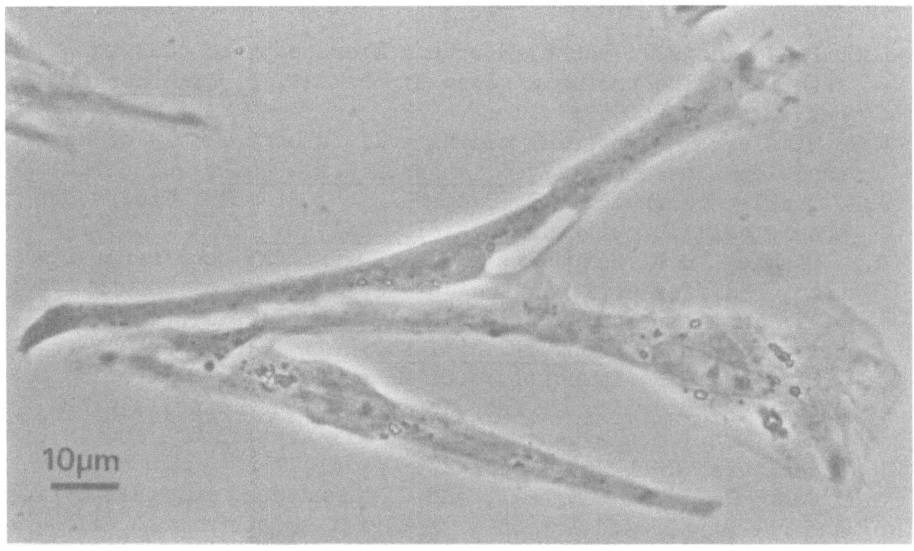

Fig. 5. Contractile state smooth muscle cells in the presence of 5% hyperlipemic whole blood serum.

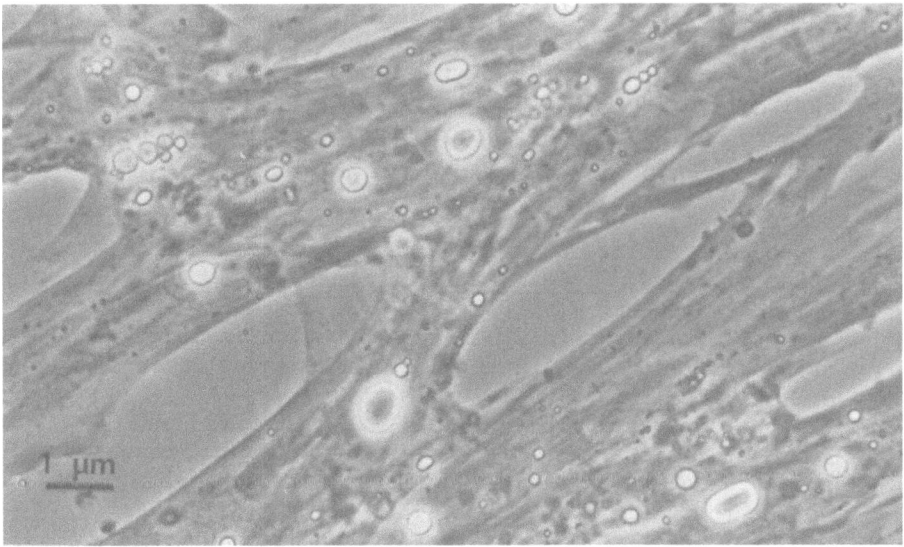

Fig. 6. Synthetic state smooth muscle cells in the presence of 5% hyperlipemic whole blood serum.

REFERENCES

1. D.C. Pease and W.J. Paule, Electron microscopy of elastic
    arteries. The thoracic aorta of the rat, J. Ultrastruct.
    Res. 3:469 (1960).
2. R.W. Wissler, The arterial medial cell, smooth muscle or multi-
    functional mesenchyme?, J. Atheroscler. Res. 8:201 (1968).
3. J.H. Chamley-Campbell, G.R. Campbell, and R. Ross, The smooth
    muscle cell in culture. Physiological Rev. 59:1 (1979).
4. J.H. Chamley, G.R. Campbell, and G. Burnstock, Dedifferentia-
    tion, redifferentiation and bundle formation of smooth muscle
    cells in tissue culture: the influence of cell number and
    nerve fibres, J. Embryol. Exp. Morphol. 32:297 (1974).
5. J.H. Chamley, G.R. Campbell, J.D. McConnell, and U. Groschel-
    Stewart, Comparison of vascular smooth muscle cells from
    adult human, monkey and rabbit in primary culture and in
    subculture, Cell Tiss. Res. 177:503 (1977).
6. G.R. Campbell, J.H. Campbell, and G. Burnstock, Differentiation
    and phenotypic modulation of arterial smooth muscle cells,
    in: "Structure and Function of the Circulation" Vol. 3,
    C.J. Schwartz, N.T. Werthessen, and S. Wolf, eds., Plenum
    Press, New York and London (1980).
7. J.C.F. Poole, S.B. Cromwell, and E.P. Benditt, Behaviour of
    smooth muscle cells and formation of extracellular struct-
    ures in the reaction of arterial walls to injury, Am. J.
    Pathol. 62:391 (1971).
8. H.C. McGill Jr., The lesion, in: "Atherosclerosis III, Proc.
    3rd Intern. Symp.", G. Schettler and A. Weizel, eds.,
    Springer-Verlag, Berlin (1974).
9. J.C. Geer, and M.D. Haust, Smooth muscle cells in atherosclerosis,
    in "Monographs on Atherosclerosis 2", O.J. Pollak, H.S. Simms
    and J.E. Kirk, eds., Karger, Basel (1972).
10. J.H. Chamley-Campbell, G.R. Campbell, and R. Ross, Phenotype-
    dependent response of cultured aortic smooth muscle to
    serum mitogens, J. Cell Biol. In press.
11. U. Groschel-Stewart, J.H. Chamley, G.R. Campbell, and G.
    Burnstock, Changes in myosin distribution in dedifferenti-
    ating and redifferentiating smooth muscle cells in tissue
    culture, Cell Tiss. Res. 165:13 (1975).
12. U. Groschel-Stewart, J.H. Chamley, J.D. McConnell and G.
    Burnstock, Comparison of the reaction of cultured smooth
    and cardiac muscle cells and fibroblasts to specific anti-
    bodies to myosin, Histochem. 43:215 (1975).
13. U. Groschel-Stewart, J. Schreiber, C. Mahlmeister, and K.
    Weber, Production of specific antibodies to contractile
    proteins, and their use in immunofluorescence microscopy.
    I. Antibodies to smooth and striated chicken muscle myosins.
    Histochem. 46:229 (1976).

14. J.H. Chamley-Campbell, and G.R. Campbell, What controls the phenotype of smooth muscle in culture? Submitted for publication.

15. R. Ross, and B. Kariya, Morphogenesis of vascular smooth muscle in atherosclerosis and cell culture, in: "Handbook of Physiology Section 2 The Cardiovascular System Vol. II Vascular Smooth Muscle," D.F. Bohr, A.P. Somlyo, and H.V. Sparks, Jr, ed., (1980).

16. D.B. Weinstein, T.E. Carew, and D. Steinberg, Uptake and degradation of low density lipoprotein by swine arterial smooth muscle cells with inhibition of cholesterol biosynthesis, Biochim. Biophys. Acta. 424:404 (1976).

17. J.H. Chamley-Campbell, G.R. Campbell, L. Popadynec, and P. Nestel, Phenotype-dependent changes in LDL metabolism of cultured smooth muscle cells, Submitted for publication.

Acknowledgement: This work was supported by the National Health and Medical Research Council of Australia and the Life Insurance Medical Research Fund of Australia and New Zealand, and was assisted by Ms. Lucy Popadynec and Ms. Janet Rogers.

THERMODYNAMICS AND MECHANISM OF THE ASSOCIATION OF PLASMA APOLIPOPROTEINS WITH SYNTHETIC PHOSPHATIDYLCHOLINES

H.J. Pownall, J.B. Massey and A.M. Gotto,Jr.

Department of Medicine
Baylor College of Medicine, Houston, Texas 77030, U.S.A.

INTRODUCTION

Two fundamental question in our investigations of lipoprotein structure and function have been how do lipoproteins form, and what factors contribute to lipoprotein stability. In the simplest sense, apolipoproteins might be considered as polymeric detergents: within lipoproteins, the apolipoproteins, phospholipids, and cholesterol form a surface monolayer, which separates a core of neutral lipids from an external aqueous compartment. In the crudest verification of lipid-protein association, one finds that most of the plasma apolipoproteins spontaneously associate with highly turbid multibilayers of phospholipid to form small lipid-protein complexes that are optically clear. In all such reassembly studies it has been found that phospholipids alone or with neutral lipids form complexes, but that the neutral lipids do not associate with apolipoproteins in the absence of phospholipids. Therefore, the phospholipid-apolipoprotein interaction would appear to be a key structural unit in the organization of a lipoprotein. It is the structure and dynamics of this association which we will attempt to correlate with what is known about the structure of lipoproteins, lipids and apoproteins.

It was noted in the early studies of lipoproteins that the structure and dynamics of lipids and protein in lipoproteins were different from those of the same components in solution (1). In dilute solution, the monomeric apolipoproteins have relatively little secondary structure; at higher concentrations the apoproteins

125

self-associate with a concurrent increase in α-helical structure.
In native and reassembled lipoproteins, the apoproteins contain even
more helical secondary structure. Therefore, the major structural
change in the apoproteins,when transferred from the aqueous phase
to a site in a lipoprotein, is that the protein becomes highly
helical. This observation led to the development of the amphipathic
helix theory of lipoproteins as first proposed by Segrest et al.(2).
According to that theory, an apolipoprotein has a distribution of
the polar and hydrophobic amino acids throughout the sequence such
that, when placed in an α-helix, they are distributed onto
opposite sides of the helix. Hypothetically, upon being transferred
from the solution to the lipid phase, the apolipoprotein becomes
more helical, with the nonpolar face of the helix penetrating into
the hydrophobic region of the lipid to about the fifth methylene unit
and the polar face being approximately coplanar with the polar head
groups of the phospholipid. When an apolipoprotein associates with
multilamellar or bilamellar arrays of phospholipids, the lipid
particles are converted into much smaller lipid–protein complexes
(3-6) in which a certain number of lipids must form a boundary with
the apoprotein. From both kinetic and calorimetric data, we have
developed a model for the mechanism of lipid–protein association and
for the stability of the resulting product.

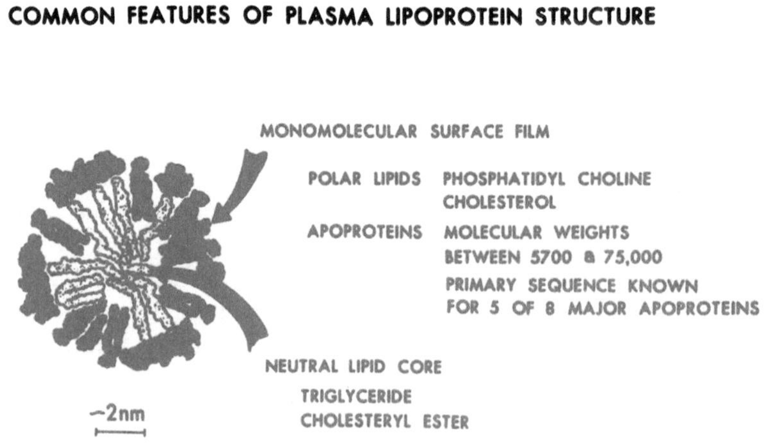

COMMON FEATURES OF PLASMA LIPOPROTEIN STRUCTURE

MONOMOLECULAR SURFACE FILM

POLAR LIPIDS   PHOSPHATIDYL CHOLINE
               CHOLESTEROL
APOPROTEINS   MOLECULAR WEIGHTS
               BETWEEN 5700 & 75,000
               PRIMARY SEQUENCE KNOWN
               FOR 5 OF 8 MAJOR APOPROTEINS

NEUTRAL LIPID CORE
TRIGLYCERIDE
CHOLESTERYL ESTER

~2nm

Fig. 1 - The structure of HDL_2 obtained from space-filling models
(adapted from Verdery and Nichols(13)). The surface contains the
polar components, apolipoprotein, phospholipid, and cholesterol
whereas the core contains the neutral lipids, mostly cholesteryl
ester.

Results and Discussion

We have measured the enthalpy of association of several
phospholipids with apolipoproteins A-II and C-III by batch calorimetry and correlated the observed enthalpy change with the difference in helical content of the starting apolipoprotein and that of
the apoprotein in the product (7). Although there is some degree of
variability, a fairly good linear relationship between the enthalpy
and helix changes is obtained. From the slope of the line in Figure 2,
we determined that there is an enthampy change of -1.3 kcal/mole of
residue of apoprotein converted from a random coil to an   α-helix.
This value is well within the range for simple model sysstems and
does not differ greatly from independently determined values for
helix formation in apoA-I (See Figure 2).

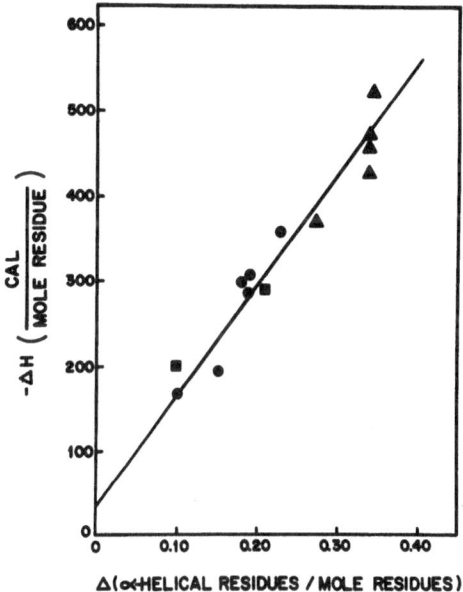

Δ(αHELICAL RESIDUES / MOLE RESIDUES)

Fig. 2 - The correlation of the enthalpy of association of apolipoproteins and phospholipids with changes in α-helicity,obtained with
a series of phosphatidylcholines and apoA-II(o), apoC-III (▲), and
apoA-I (■)(14) (□) (23). The curve is a linear regression fit to the
data with a slope corresponding to -1.3 kcal/helical residue formed,
an x-intercept of 39 cal, and a correlation coefficient of 0.94.
The data for apoA-I were not used in this calculation since they were
not obtained under the same conditions as those for apoA-II and
apoC-III association with lipid.

   In addition to the changes in the protein structure, there are
temperature-dependent changes in the phospholipid structure, which
accompany lipid-protein complex formation. The association of
apoA-II and dimyristoylphosphatidylcholine (DMPC) is well-suited for
thermodynamic and mechanistic studies since their association is
spontaneous, occurs aver a period of minutes, and is quantitative at
initial lipid-protein ratios whose values are temperature dependent.
Furthermore, DMPC is a well-characterized phospholipid having an
order (gel) → disorder (liquid crystalline) transition of its acyl
chains at an experimentally convenient temperature, T_c = 23.9° (8).
By contrast, the T_c of the products formed by the association of
DMPC and apoA-II are somewhat higher, ~27° . The composition and
structure of the products are a function of the temperature at which
the reaction occurs. Distinct products form below the T_c of DMPC
(T < 23.9°), above the T_c of the resulting complex (T > 27°) and between
the T_c of DMPC and the lipid-protein complex (23.9°- ~ 27°). Using
differential scanning calorimetric (DSC) data we determined the
amount of boundary lipid around each apoA-II in these three complexes; these values and some of the properties of the complexes are
summarized in Table I .

Table I

Properties of the Complexes of apoA-II and DMPC Formed in Three
Temperature Regions

	apoA-II	30° (T > T_c complex)	24° or 29° (T < T_c complex)	24° T < T_c complex, T = T_c DMPC
Stoichiometry	—	45/1	75/1[*]	240/1[*]
Molecular Weight	—	0.23×10^6	0.34×10^6	1.58×10^6
Boundary Lipid	—	38	45	45
Helical content	33%	47%	70%	70%

[*] At 24°, the 75/1 complex is formed initially so that this complex
is a precursor to the 240/1 complex.

From these data and the known enthalpies of the changes in lipid and
protein structure, it is possible to calculate the enthalpy of
formation of a DMPC:apoA-II complex as a function of temperature.

The model upon which the calculation is based is given in Figure 3.
The conversion of apoA-II from a random coil to an α-helix is the
major enthalpic contribution of the changes in protein structure to
the total enthalpy of association. We use the directly determined
value for this process; this is −2.9 kcal/1% conversion of apoA-II
from a random coil to an α-helix. The enthalpic contribution of
changes in lipid structure are based upon the physical state of the
reactants (liposomes) and the product (complex). Hypothetically,
the three physical states of the lipid in this system are gel,
liquid crystalline and boundary. The transfer of DMPC from the
boundary or gel phase to the liquid crystalline phase has an
enthalpy of + 5.4 kcal/mole.

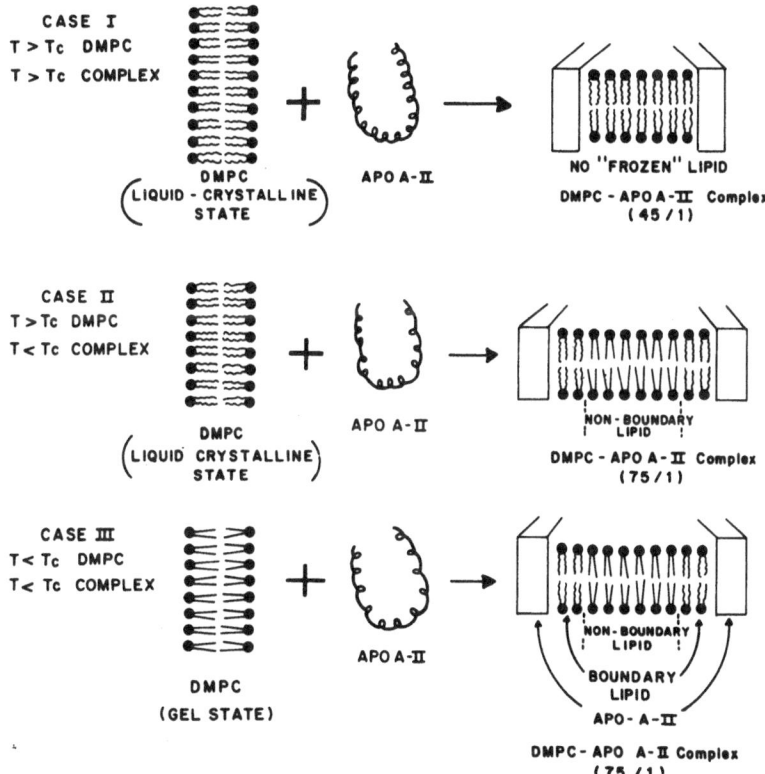

Fig. 3 − A schematic model that relates the enthalpic contributions
of structural changes in DMPC (freezing above the T_c of DMPC and
below the T_c of complex) and formation of "boundary" lipid below T_c
of DMPC and apoA-II (helix formation) to the total enthalpy of
complex formation. The complexes are represented as disks with
protein on the edge.

Therefore, the changes in the lipid structure wich accompany
lipid-protein association are strongly temperature dependent. Above
the T_c of the complex and of DMPC there is no change in the physical
state of the lipid upon formation of a complex with apoA-II. The
calculated enthalpy of association is due only to helix formation
and is -58 kcal/mole apoA-II. The experimental value is -62 kcal/
mole apoA-II. Between the transition temperature of DMPC and that
of the complex the reaction is predicted to be highly exothermic due
to helix formation plus the conversion of 30 moles of "nonboundary"
DMPC from the disordered to ordered state. The sum of these is
-289 kcal/mole apoA-II. The experimental value is -260 kcal/mole
apoA-II. Finally, below T_c, apoA-II converts 45 "ordered" lipid
molecules to boundary lipid molecules with a concomitant 42% increase in helical content. The respective calculated and observed
values are + 121 and + 90 kcal/mole apoA-II.

In spite of the very large positive enthalpy of association,
below T_c the reaction occurs spontaneously so that for this reac-
tion, $\Delta G < 0$. Since $\Delta G = \Delta H - T\Delta S$, then $T\Delta S > \Delta H$ and the reaction is
entropically driven. The most likely source of the change in
entropy is the transfer of nonpolar amino acid side chains from the
aqueous phase to the hydrocarbon region of the lipid. Given the
stability of the product, ΔG must be <-10 kcal/mole apoA-II, giving
a calculated value for $T\Delta S$ of >100 kcal/mole apoA-II. This value is
similar to that calculated from the sum of the free energies of
transfer of the hydrophobic side chains of the amphipathic helical
regions of apoA-II from an aqueous to a hydrocarbon environment.
Using the hydrophobicity values of Bull and Breese (9), the calculated free energy of transfer of the side chains is -98 kcal/mole
of apoA-II.

Thus we have identified three significant terms which contribute to the stability of a lipid-protein complex. These are helix
formation (exothermic), formation of boundary lipid (endothermic
below T_c) or crystallization of nonboundary lipid (exothermic) and
transfer of nonpolar residues from water to the hydrocarbon region
of the phospholipid ($T\Delta S > 100$ kcal).

Kinetics of Lipid-Protein Complex Formation

We have studied the rate of association of apoA-I (5) and
apoA-II (10) with DMPC in some detail. ApoA-I and apoA-II associate

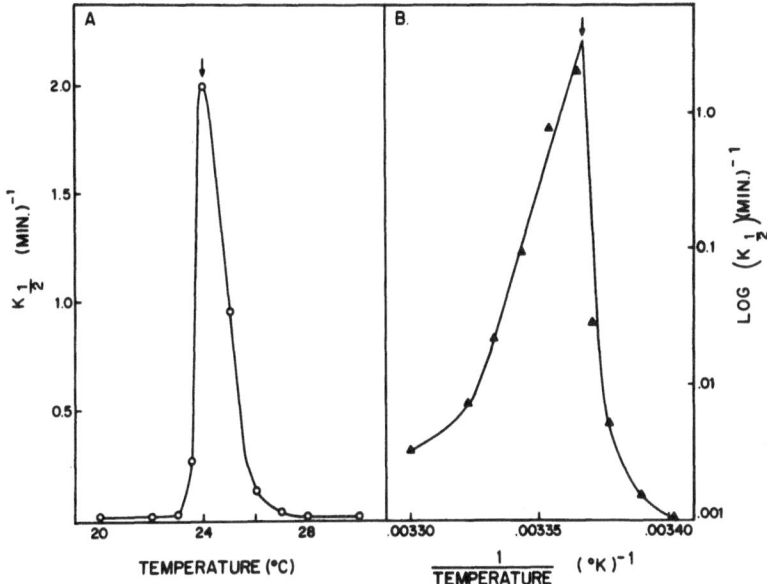

Fig. 4 - Temperature dependence of the rate of association of apoA-I with DMPC: (A) linear plot; (B) Arrhenius plot showing maximum at T_c.

with multilamellar liposomes of DMPC with a concomitant clarification of the lipid turbidity. The rate of disappearance of the turbidity can be used to estimate the rate of complex formation of DMPC with apoA-I and apoA-II. The rate of complex formation of apoA-I with DMPC has an unusual temperature dependence, which is shown in Figure 4. Above and below the T_c of DMPC, the rate is very slow but increases by several orders of magnitude at T_c. Since the rates of reaction of apoA-I with the ordered and disordered phases of DMPC are relatively slow, there must be some unique property of coexisting phases that facilitates lipid-protein association. We believe this property is a higher permeability due to the existence of "hole" or "channel" defects that form at the borders of the ordered and disordered phases.

It is known that the addition of impurities to pure crystals increases the number of defects. By analogy, in a DMPC matrix addition of an impurity might be expected to increase the number of defects and thereby increase the rate of reaction (11). A physiologically important "impurity" is cholesterol and the effect of temperature and cholesterol content on the rate of apoA-I-DMPC association is shown in Figure 5. At each temperature the rate is

fastest at 12.5 mole% cholesterol and, similar to our data without
cholesterol, the rates at each cholesterol content are fastest at T_c.
The combined effects of temperature and cholesterol content change
the reaction rate by three orders of magnitude. Since it is known
that DMPC and cholesterol form a separate phase containing 25 mole%
cholesterol (12), we have proposed a model for the association of
apoA-I with DMPC:cholesterol mixtures, which is shown in Figure 6.

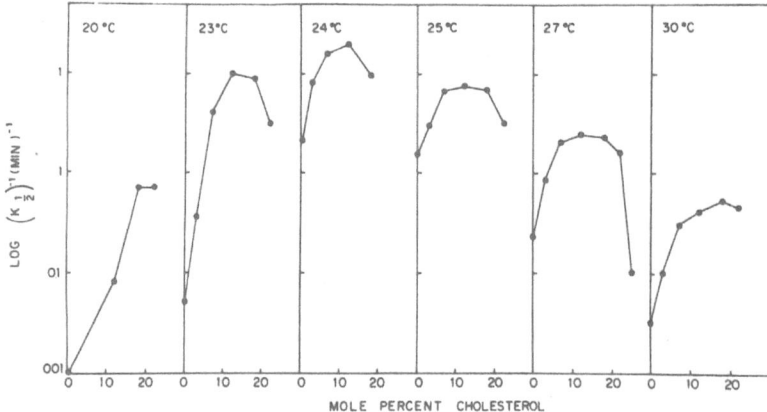

Fig. 5 - Effect of cholesterol content and temperature on the rate
of association of apoA-I and DMPC .

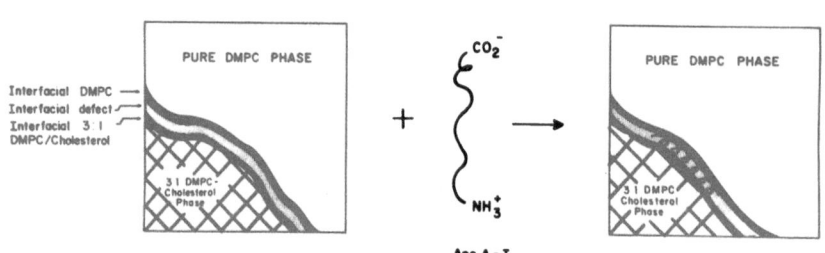

Fig. 6 - A hypothetical model for the mechanism by which apoA-I associates with a lipid matrix. On the left is a section of the surface
of a liposome in which a DMPC and a 1:3 cholesterol:DMPC phase coexist.
Each phase is sounded by interfacial lipid and the interfacial lipid
of each phase is separated from that of the other phase by a hole
or channel defect. ApoA-I may insert into this defect to give the
initial lipid-protein intermediate on the right. We have drawn the
apolipoprotein as a helical structure since it probably exists as
such in HDL and in apoA-I:PC complexes.

We suggest that pure DMPC exists in equilibrium with a 3:1 DMPC:cholesterol phase containing 25 mole% cholesterol. Between these two phases there are large defects into which the apoprotein inserts. At all sterol concentrations the rate is fastest at T_c so a relatively pure DMPC phase must be involved in the rate limiting step. If we assume that the domain sizes of the pure DMPC and the 3:1 DMPC:cholesterol phases are equal then the maximum number of hole defects and, consequently, the maximum rate of reaction should occur when the number of moles in the two phases is equal. This is calculated to be 12.5 mole%.

The temperature dependence of the rate of apoA-II:DMPC association is, in some ways, similar to that described for apoA-I but has the additional complication of forming multiple products whose structure is also a function of reaction temperature. In Figure 7A, we show, for the sake of comparison, the temperature dependence of the rate of association of apoA-II with DMPC at two different lipid to protein ratios. As estimated from the turbidity change (Figure 7B) the reaction of DMPC and apoA-II goes to completion at low DMPC: apoA-II ratios. At low DMPC to apoA-II ratios the behaviour is similar to that of apoA-I; the rate is fastest at T_c but the temperature range over which the reaction occurs is considerably broader than that of apoA-I (compare Figure 4A and 7A). By contrast, at high lipid to protein ratios, the extent of clarification of turbidity (Figure 7B) indicates quantitative association only at T_c; however, the rate of association (Figure 7A) at high DMPC to apoA-II ratios is at a significantly higher temperature than the T_c of DMPC; however, the temperature at which the rate is fastest coincides with the T_c ($\sim 27°$) of an apoA-II:DMPC complex. Therefore, we speculate that at high DMPC concentrations an initial lipid-protein complex is formed and that the association of additional lipid with the initially formed complex may depend upon the coexistence of ordered and disordered lipid phases within the precursor complex. This is predicted to be at the T_c of the complex ($\sim 27°$) rather than that of DMPC ($23.9°$).

Summary of the Proposed Kinetic and Thermodynamic Events Leading to Lipid-Apolipoprotein Complex Formation

Although more evidence is needed for a complete understanding of how and why apolipoproteins and phospholipids associate, we have developed a working model describing the sequence of events leading

to the formation of a lipid-apolipoprotein complex. The general
scheme shown in Figure 8 is based upond data given here as well as
published work from other laboratories.

Hypothetically, the apolipoprotein can exist in many conformational states, such as lipid-free monomers in rapid equilibrium with
oligomers or with monomeric lipid bound to high affinity sites on
the apolipoprotein. In both cases new α-helical regions are formed;
this is an exothermic process with H = − 1.3 < 0.3 kcal/mole of
α-helical residues formed. The apolipoprotein binds to bulk phase
lipid which may be in the ordered phase ($T < T_c$), disordered phase
($T < T_c$) or in coexisting ordered and disordered phases ($T = T_c$).
Our kinetic data suggest that reaction in the latter case is most
probable and that the precursor lipid-protein complex preferentially
penetrates into a preexisting "hole" at T_c. As the apolipoprotein
enters the "hole" the apolipoprotein folds into an amphipathic helix
with the hydrophilic face exposed to the aqueous phase and the
hydrophobic face penetrating part way into the hydrocarbon region.

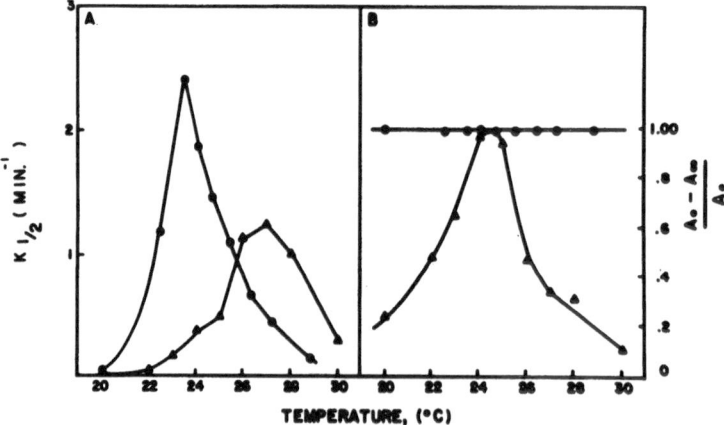

Fig. 7 - (A) The temperature dependence of the rate of disappearance
of DMPC liposomal turbidity ($k_{1/2}$) after the addition of apoA-II.
The $k_{1/2}$ values are for molar ratios of DMPC to apoA-II for 45 to 1
(o) and 100 to 1 (Δ). The insert shows the effect of increasing
reactant concentrations on the rate of clearing of a constant molar
ratio (50/1) of DMPC and apoA-II. (B) The temperature dependence of
the amount of turbidity change ($A_O − A_\infty$)/A_O at 45/1 (o) and 100/1
(Δ) molar ratios of DMPC to apoA-II.

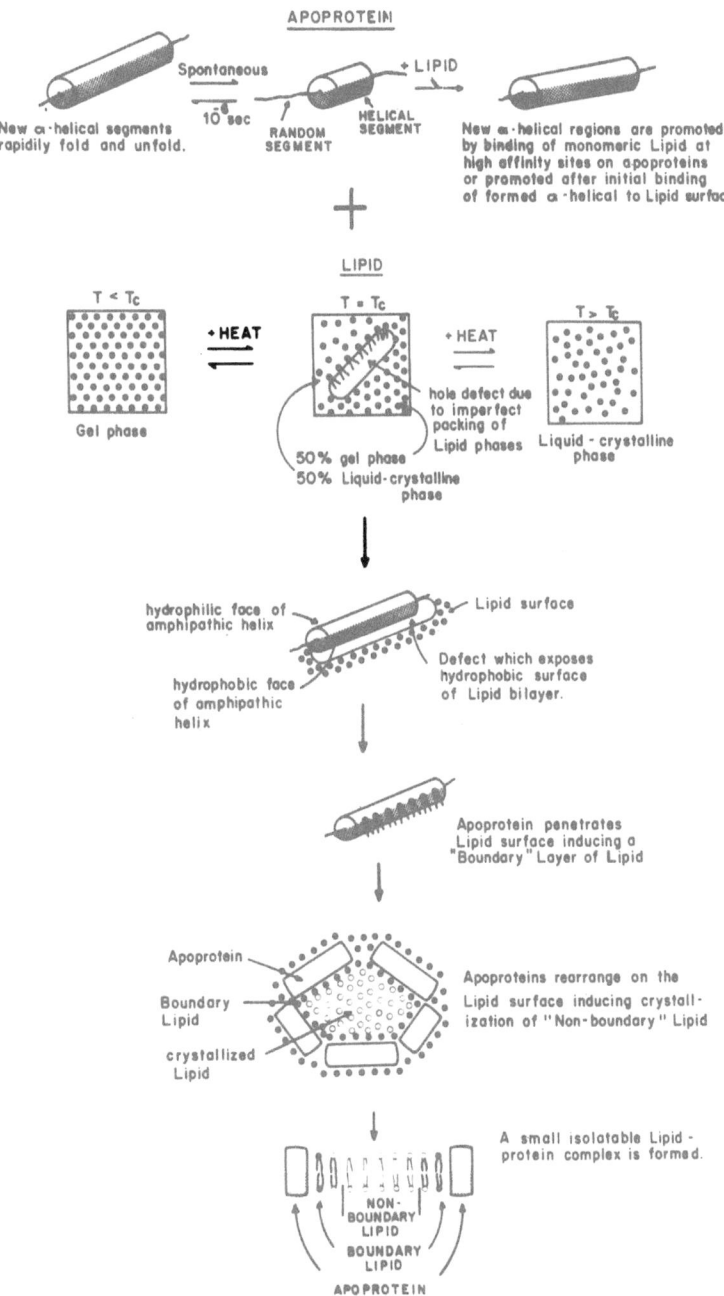

Fig. 8 – Summary of proposed thermodynamic and kinetic events involved in the association of a soluble plasma apolipoprotein with a phospholipid surface.

The $-T\Delta S$ contribution to the latter process may be calculated from
the sum of the free energy of transfer of the individual hydrophobic
amino acids on the hydrophobic face.

Within the lipid matrix, the apolipoprotein associates with a
boundary layer of phospholipid. The enthalpy for this step depends
upon the temperature with respect to the melting point of the lipid.
Below T_c the enthalpy for this step is endothermic; at and above T_c
the enthalpy is zero. Finally, we speculate that the apoprotein
rearranges on the surface crystallizing "nonboundary" lipid (highly
exothermic) to form an unstable lipid-protein domain, which dis-
sociates away from the surface in the form of a bilayer disk.

This mechanism may be important to the transfer of phospholipids and proteins to HDL during the lipolysis of triglyceride-rich
lipoproteins. We anticipate that a better understanding of the
structure and stability of lipoproteins will continue to evolve
from additional physical studies of lipid-protein reassembly.

Acknowledgement

This research was developed by the Section on Atherosclerosis,
Lipids and Lipoproteins of the National Heart and Blood Vessel
Research and Demonstration Center, a grant-supported research
project of the National Heart, Lung and Blood Institute,National
Institutes of Health, Grant No. HL-17269. HJP is an Established
Investigator of the American Heart Association.

References

1. J.D. Morrisett,R.L. Jackson, and A.M. Gotto,Jr., Lipoproteins:
   Structure and Function. Ann.Rev.Biochem. 44:183 (1975)
2. J.P. Segrest,R.L. Jackson,J.D. Morrisett, and A.M. Gotto,Jr.,
   A Molecular Theory of Lipid-Protein Interactions in the Plasma
   Lipoproteins. FEBS Lett. 38:247 (1974)
3. M.C. Ritter and A.M. Scanu, Role of Apolipoprotein A-I in the
   Structure of Human Serum High Density Lipoproteins. J.Biol.Chem.
   252:1208 (1977)
4. A.R. Tall,D.M. Small,R.J. Deckelbaum, and G.G. Shipley, Structure
   and Thermodynamic Properties of High Density Lipoprotein Re-
   combinants. J.Biol.Chem. 252:4701 (1977)
5. H.J. Pownall,J.B. Massey,S.K. Kusserow, and A.M. Gotto,Jr.,
   Kinetics of Lipid-Protein Interactions: Interaction of Apolipo-

protein A-I from Human Plasma High Density Lipoproteins with Phosphatidylcholines. Biochemistry 17:1183 (1978)

6. A. Jonas,D.J. Krajnovich, and B.W. Patterson, Physical Properties of Isolated Complexes of Human and Bovine A-I Apolipoproteins with L-α-Dimyristoyl Phosphatidylcholine. J.Biol.Chem. 252:2200 (1977)

7. J.B. Massey,A.M. Gotto,Jr., and H.J. Pownall, Contribution of α-Helix Formation in Human Plasma Apolipoproteins to their Enthalpy of Association with Phospholipids. J.Biol.Chem. 254:9359 (1979)

8. S. Mabrey and J.M. Sturtevant, Investigation of Phase Transitions of Lipids and Lipid Mixtures by High Sensitivity Differential Scanning Calorimetry. Proc.Natl.Acad.Sci. 73:3862 (1976)

9. H.B. Bull and K. Breese, Surface Tension of Amino Acid Solutions: A Hydrophobicity Scale of the Amino Acid Residues. Arch.Biochem. Biophys. 161:665 (1974)

10. H.J. Pownall,J.B. Massey,J.T. Sparrow, and A.M. Gotto,Jr., Kinetics and Mechanism of Lipid-Apolipoprotein Association. in "Proc. 25th Colloquium of the Protides of the Biological Fluids", H.Peeters ed.,Pergamon Press (1978)

11. H.J. Pownall,J.B. Massey,S.K. Kusserow, and A.M. Gotto,Jr., Kinetics of Lipid-Protein Interactions: Effect of Cholesterol on the Association of Human Plasma High Density Apolipoprotein A-I with L-α-Dimyristoylphosphatidylcholine. Biochemistry 18:574 (1979)

12. T.N. Estep,D.B. Mountcastle,R.L. Biltonen, and T.E. Thompson, Studies on the Anomalous Thermotropic Behavior of Aqueous Dispersions of Dipalmitoylphosphatidylcholine-Cholesterol Mixtures. Biochemistry 17:1974 (1978)

METABOLIC FATE OF ^{3}H-CHOLESTERYL LINOLEYL ETHER,

A NONDEGRADABLE ANALOG OF LIPOPROTEIN CHOLESTERYL ESTER

Yechezkiel Stein, Olga Stein and Gideon Halperin

Lipid Research Laboratory, Department of Medicine B,
Hadassah University Hospital, and Department of
Experimental Medicine and Cancer Research,
Hebrew University-Hadassah Medical School,
Jerusalem (Israel)

INTRODUCTION

Studies in the past have pointed out that in the human during the intravascular metabolism of very low density lipoproteins (VLDL) which culminates in the formation of low density lipoproteins (LDL) there is a considerable loss of cholesteryl ester (1). To follow the fate of cholesteryl ester during the metabolism of VLDL, various approaches were used to label the lipid portion of the lipoprotein. These can be divided into two main categories, i.e., endogenous labeling and exogenous labeling. The first was based on the injection of crystalline suspensions of labeled cholesterol of high specific activity which following uptake by the liver reappears in the VLDL secreted into plasma in the form of free and esterified cholesterol. One can remove the free cholesterol by exchange following incubation with erythrocytes, but this procedure changes the biological behavior of the lipoprotein (2) and also leaves a considerable portion of unesterified labeled cholesterol. Another approach utilizes the LCAT reaction in vitro to convert exogenously added labeled free cholesterol to cholesteryl ester, but this reaction as well does not go to completion. Therefore, several methods were developed to label lipoproteins with pure labeled cholesteryl ester by the introduction of the lipid into the lipoprotein in vitro (3). As cholesteryl ester is a very nonpolar compound various solvents with detergent properties were used to this end (4). The disadvantage of such an approach is that the detergent at concentrations used is deleterious to the lipoprotein particle and shortens its half life in the circulation. Apart from this disadvantage there remains the major problem, namely that cholesteryl ester following its uptake by cells is easily degradable (5). This makes it difficult to distinguish from free cholesterol which may arrive at a certain cell, tissue or organ by a route and mechanism not applicable to cholesteryl ester. One example of a problem which is extremely difficult to solve using lipoproteins labeled with cholesteryl ester is the quantitation of the participation of various organs in the uptake of a given lipoprotein cholesteryl ester. These considerations have prompted us to search for a nondegradable analog of cholesteryl ester and for a method which would permit efficient labeling of various lipoproteins and which would not have deleterious effects on their biological behavior.

METHODOLOGY

Synthesis of Cholesteryl Linoleyl Ether

Synthesis of cholesteryl oleyl or linoleyl ether was carried out by a modification (6) of the method of Stoll (7). Briefly, $[7\alpha(n)-^3H]$cholesterol (0.5-1 mCi, specific activity 9.5 Ci/mmol) was reacted with p-toluenesulfonyl chloride in dry pyridine at 38°C for 20 h. Following extraction with hexane, the cholesterol toluene sulfonate was transferred to ampules, the hexane evaporated, a 20-50-fold excess of linoleyl alcohol added and the ampule sealed under N_2 and heated for 2 h at 110°C. The reaction mixture was extracted with hexane in the presence of excess $NaHCO_3$ and the cholesteryl alkyl ether was chromatographed on a silicic acid column using hexane and increasing concentrations of benzene as eluent. The labeled cholesteryl alkyl ether was eluted in 10-15% benzene in hexane. The purity of the compounds was ascertained by thin-layer chromatography (3% diethyl ether in light petroleum, b.p. 30-60°C). More than 98% of the ^3H label was recovered in the region of the cholesteryl ether. On thin-layer chromatography the R_f of the cholesteryl alkyl ethers was similar to that of cholesteryl linoleate (8). The mass spectrum of cholesteryl linoleyl ether is shown in Fig. 1. The parent peak has the expected mass number of 634 and the fragmentation of ring A between C17 and C20 gives the expected mass number of 521. The cleavage of the alkadienyloxy moiety with or without hydrogen shift gives the expected m/e of 368, 369 and 370. $[^{14}C]$Cholesteryl linoleate was prepared and purified as described before (9).

Biotransfer of Labeled Cholesteryl Linoleyl Ether to
Intralipid, VLDL and LDL

To introduce the labeled cholesteryl linoleyl ether into lipoproteins several approaches were tried and discarded. Optimal results were obtained by the combination of the following methodology. Human or rat high density lipoprotein (HDL) was partially delipidated using cold heptane (3). The labeled ^3H-cholesteryl linoleyl ether and ^{14}C-cholesteryl linoleate were dried from heptane, suspended in 0.15 M NaCl by sonication for 30". The lipid sonicate

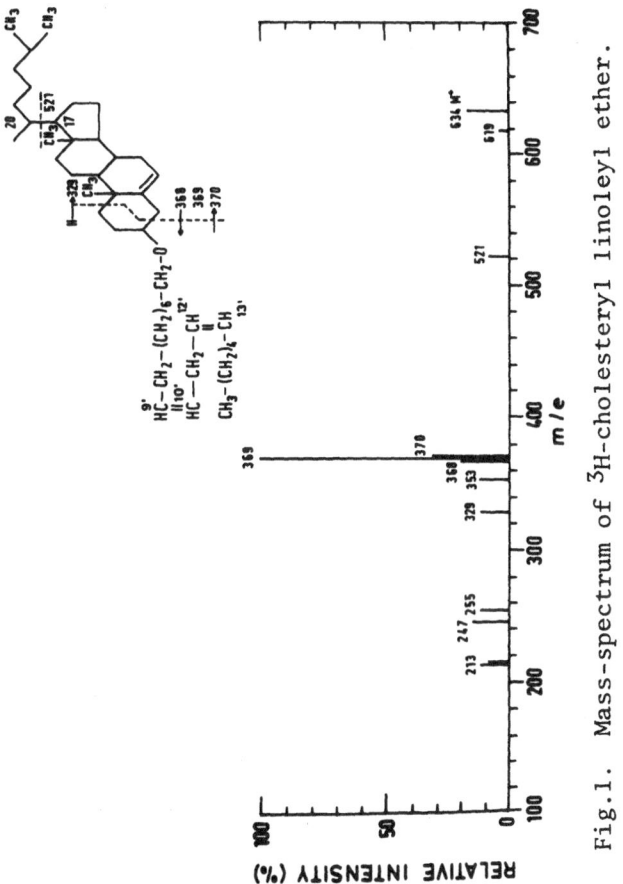

Fig.1. Mass-spectrum of ³H-cholesteryl linoleyl ether.

was then cosonicated twice for 30" with the delipidated HDL dissolved
in 0.15 M NaCl in the presence of free cholesterol (4). The labeled
relipidated "HDL" was separated from loosely bound lipid by centrifugation for 60 min in a SW41 rotor at 39000 RPM. About 50-70% of
the original radioactivity of both ^{3}H-cholesteryl linoleyl ether
and ^{14}C-cholesteryl linoleate were recovered with the "HDL" fraction.
This "HDL" was used as a source of the labeled lipid to be transferred to VLDL or Intralipid. For the labeling of LDL the "HDL"
was cleared further by centrifugation at d = 1.063 for 24 h, at
39000 RPM in SW41 rotor. The transfer of the label from "HDL" to
the other lipoproteins was carried out by incubation in the presence
of d > 1.25 fractions of human serum, which contains the cholesteryl
ester transfer protein (10-12). The labeled Intralipid or VLDL were
isolated by centrifugation for 60 min or 24 h in SW41 rotor at
39000 RPM; LDL was reisolated following a 24 h spin at d = 1.063.
The yield of radioactivity of both labels with all lipid acceptors
used was 50% or more.

RESULTS

The first experiments were designed to learn about the chemical
composition, electrophoretic mobility and EM appearance of VLDL and
LDL labeled by the biotransfer procedure. Following reisolation
from the labeling mixture both lipoproteins retained their chemical
composition; on agarose gel electrophoresis the label was shown
to comigrate with the protein peak of VLDL and no change was seen
in size and shape of VLDL and LDL particles examined in negatively
stained preparations. To test the nondegradability of the labeled
cholesteryl linoleyl ether by cellular enzymes, labeled LDL was
introduced into the culture medium of human skin fibroblasts. While
98% of intracellular labeled cholesteryl linoleate had been degraded
after 4 days of incubation, 92% of intracellular cholesteryl linoleyl
ether remained intact(8).

In another set of experiments, the transfer of cholesteryl
linoleyl ether from "HDL" to Intralipid or LDL was compared. As
seen in Table I, not more than 1% of the label floated during ultracentrifugation in the absence of an acceptor, i.e., Intralipid or
VLDL. The transfer of both labeled cholesteryl linoleyl ether and
cholesteryl ester was low when incubation was carried out at 4°C in
the presence of either human or rat d > 1.25 g/ml fraction of serum.
A pronounced and comparable transfer of cholesteryl linoleyl ether
and cholesteryl ester was obtained in the presence of human d >
1.25 g/ml fraction, with Intralipid or VLDL as acceptors. However,
only minimal transfer of labeled cholesteryl linoleyl ether occurred when human d > 1.25 g/ml fraction was replaced by rat
d > 1.25 g/ml fraction. The comparative transfer of cholesteryl
linoleyl ether and cholesteryl ester from human HDL to Intralipid
was further studied by varying the time of incubation, the concentration of the d > 1.25 g/ml fraction and of the acceptor Intralipid

Table 1. Effect of Temperature and the Source of d > 1.25 g/ml
Fraction on Transfer of [^{3}H]Cholesteryl Linoleyl Ether
and [^{14}C]Cholesteryl linoleate from Rat or Human HDL to
Intralipid or VLDL

The incubation mixture contained Intralipid or VLDL, labeled with
[^{3}H]cholesteryl linoleyl ether and [^{14}C]cholesteryl linoleate in
360 μg HDL protein and d > 1.25 g/ml fraction of serum to give 60 mg
protein/ml, final volume 0.7 ml. At the end of 24 h incubation, the
mixture was overlayered with 10 ml saline and centrifuged for 60 min
or 24 h for Intralipid and VLDL respectively, at 39000 RPM in
SW41 rotor.

Acceptor ·(mg TG)	d > 1.25 g/ml fraction of serum	Temper- ature (°C)	Label recovered (%) in Intralipid or VLDL	
			^{3}H-cho- lesteryl ether	^{14}C-cho- lesteryl ester
None	Human	37	1	1
Intralipid (5)	Human	4	16	16
Intralipid (5)	Human	37	84	84
Intralipid (5)	Rat	37	7	10
Human VLDL (2.6)	Human	37	59	64
Human VLDL (2.6)	None	37	21	26
Rat VLDL (0.5)	Human	37	65	65
Rat VLDL (0.5)	Human	4	25	22
Rat VLDL (0.5)	None	37	25	20

Adapted from Stein et al., Biochim.Biophys.Acta 620 (1980), 247-260
(4).

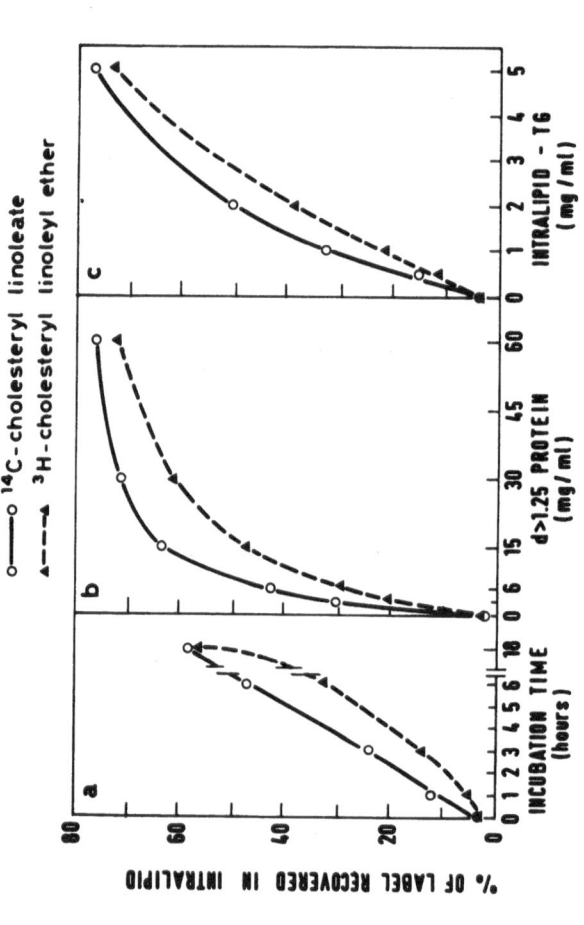

Fig.2.   Transfer of [³H]cholesteryl linoleyl ether and [¹⁴C]cholesteryl linoleate from HDL
to Intralipid. (a) Incubation conditions: Intralipid triacylglycerol 5 mg/ml, human
HDL 1 mg protein/ml, human d>1.25, 30 mg protein/ml; (b) the concentration of Intralipid and HDL as in (a), incubation time 18 h; (c) the concentration of HDL and d >
1.25 fraction same as in (a), and incubation time 18 h. All incubations were carried
out at 37°C in SW41 tubes. At the end the incubation mixture (1-1.5 ml) was overlayered with saline, and the labeled Intralipid isolated by centrifugation. (From Stein
et al., Biochim.Biophys.Acta, 260 (1980), 247-260).

(Fig. 2). It seems that the transfer of the labeled cholesteryl ester progressed at a somewhat more rapid initial rate than that of the cholesteryl linoleyl ether but by extending the time of incubation a similar percentage of both labeled compounds was recovered in the Intralipid. The transfer of the labeled cholesteryl linoleyl ether lagged behind the transfer of the cholesteryl ester at low concentrations of d > 1.25 g/ml fraction of serum (panel b) but this difference disappeared at high protein concentrations. The transfer of both labeled compounds increased with the amount of Intralipid added and reached 75% in the presence of 5 mg/ml Intralipid triacylglycerol (panel c).

The biological behavior of the VLDL labeled with ³H-cholesteryl linoleyl ether was studied following intravenous injection into rats. As seen in Table II, about 90% of the injected dose was recovered in the circulation 5 min after injection and during the first 10 min the liver and plasma accounted for more than 98% of the injected dose. The tissue distribution of ³H-cholesteryl linoleyl ether and ¹⁴C-cholesteryl linoleate was studied at later time intervals, when plasma label has been maximally reduced. The data summarized in Table III show that almost 90% of the VLDL derived ³H-cholesteryl linoleyl ether is found in the liver between 3 and 48 h after injection, while only 23% of ¹⁴C-cholesteryl ester remained in the liver after 6 h. More than 85% of the ¹⁴C label was recovered in free cholesterol, whereas more than 90% of ³H-cholesteryl linoleyl ether had not been hydrolyzed even 48 h after injection. Using radioautography on frozen sections, the uptake of the ³H-cholesteryl linoleyl ether could be localized to hepatocytes (Fig. 3), which in previous studies were shown to be the main site of uptake of ¹²⁵I-labeled VLDL (13). Up to 4.0% of injected labeled cholesteryl linoleyl ether was found in the whole muscle mass and up to 0.4% in adrenals.

In the next experiments, the LDL labeled with ³H-cholesteryl linoleyl ether and ¹⁴C-cholesteryl linoleate was studied after injection into rats (14). As seen in Fig. 4, the disappearance of the labeled lipids from the circulation was measured and the clearance of ³H-cholesteryl linoleyl ether and ¹⁴C-cholesteryl linoleate was equal. The t½ of both labels in the circulation was about 7 h and up to 8 h after injection more than 94% of the ³H-radioactivity in plasma was precipitated with heparin-manganese, indicating that no transfer of the label from LDL to HDL had occurred. In contradistinction to the findings with VLDL, the uptake of the ³H-labeled cholesteryl linoleyl ether by the liver was much slower and even after 24 h it was only about 30% of the injected dose (Table IV). The tissue distribution of the ³H-cholesteryl linoleyl ether is shown in Table V and it can be seen that in addition to the liver, the small intestine and the carcass (muscle, fat and skin) are the important sites of uptake of LDL derived ³H-cholesteryl linoleyl ether. So far, only male rats were used and in the following

Table II. Clearance of VLDL Labeled with ^{3}H-Cholesteryl Linoleyl
           Ether from the Circulation

Six male rats, 180-200 g of body weight, were injected into the
tail vein with rat VLDL (0.25 mg TG). After given time intervals
the rats were killed under ether anaesthesia. Values are means ± S.E.

| | % of injected dose/organ | |
	5 min	10 min
Liver	28.2 ± 3.4	58.7 ± 5.7
Plasma	69.1 ± 4.0	40.2 ± 3.2

Table III. Distribution of Label after Injection of VLDL Labeled
            with [3H]Cholesteryl Linoleyl Ether and [14C]Cholesteryl
            linoleate

The rats (160 g body wt.) were injected with homologous VLDL (0.3-
1.0 mg TG/rat) labeled with [3H]cholesteryl linoleyl ether and with
[14C]cholesteryl linoleate. Values are means of label recovered in
whole organ. Plasma volume and muscle mass were estimated as 4 and
45% of body weight, respectively. Number of rats, in parentheses.

| Organs | Injected dose recovered (%) after injection of VLDL labeled with: | | | | |
| | [3H]Cholesteryl linoleyl ether | | | [14C]Cholesteryl linoleate | |
h:	3 (4)	6 (5)	48 (2)	3 (3)	6 (3)
Plasma	2.2	2.5	0.6	5.9	6.5
Liver	88.3	84.8	89.5	35.1	23.9
Spleen	0.7	0.7	1.5	1.2	1.6
Lung	0.2	0.3	0.6	1.4	2.0
Heart	0.06	0.1	-	0.2	0.4
Muscle	1.4	-	4.0	4.8	10.3
Adrenal	0.2	0.2	0.4	0.2	0.1

Adapted from Stein et al., Biochim.Biophys.Acta,260 (1980),247-260.

Table IV. Uptake of ^{3}H-Cholesteryl Linoleyl Ether and
^{14}C-Cholesteryl Linoleate Labeled LDL by Rat Liver

Male rats, 160-180 g b.wt., were injected with up to 260 μg of
human LDL protein labeled with ^{3}H-cholesteryl linoleyl ether and
^{14}C-cholesteryl linoleate. For each time interval 3-11 rats were
used. Values are means ± S.E.

Time, h	^{3}H	^{14}C
	% injected dose/liver	
0.25	5.0 ± 0.5	3.9 ± 0.3
1	10.7 ± 0.6	6.5 ± 0.1
4	14.2 ± 0.7	7.5 ± 0.9
8	22.2 ± 2.0	11.0 ± 0.7
24	28.8 ± 2.5	8.9 ± 0.2

Adapted from Stein et al., Biochim.Biophys.Acta. 1980 (14).

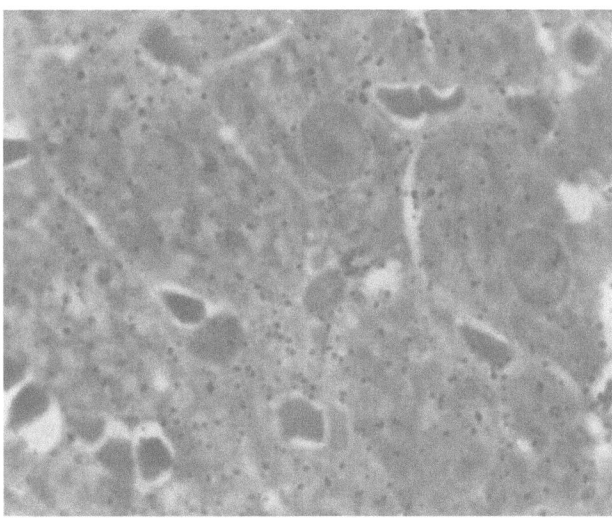

Fig.3. Autoradiograph of frozen section of rat
liver 6 h after injection of VLDL labeled
with ^{3}H-cholesteryl linoleyl ether.
x 1500. From Stein et al., Biochim.Biophys.
Acta, 1980 (4).

Table V.   Recovery of ^{3}H-Cholesteryl Linoleyl Ether after
           Injection of Labeled LDL

Male rats were injected with 80 μg of human LDL protein labeled with
[3]H-cholesteryl linoleyl ether. The rats were killed 24 h after injection. The radioactivity in the carcass was determined after
alkaline hydrolysis and extraction of the labeled lipid with petroleum ether. The values are means ±S.E. of 4 rats.

	% injected dose/organ
Plasma	6.8 ± 0.4
Liver	32.9 ± 3.1
Small intestine	6.1 ± 0.2
Carcass	34.7 ± 1.3
Other	7.6 ± 0.7
Total recovery	88.1

experiments, the uptake of ^{3}H-cholesteryl linoleyl ether was compared in male and female rats injected with the labeled LDL. The
data shown in Table VI are from male and female rats, 2 and 6 months
of age. In the two-month-old female rats the uptake of the [3]H-
cholesteryl linoleyl ether by the liver and by the adrenals was
slightly higher than in the males. However, there was no difference in the older age group. A significant uptake of up to 0.6%
of injected dose was recovered in the gonads of both species; the
ovaries being much smaller than the testes, had a higher specific
activity than the latter. This small difference in the uptake of
[3]H-cholesteryl linoleyl ether by the liver and adrenals after injection into young male and female rats could be very markedly
enhanced by treatment of male rats with 17α-ethinyl estradiol for
5 days. As seen in Table VII, more than 80% of injected ^{3}H-cho-
lesteryl linoleyl ether were recovered in the liver of the estradiol-
treated rats, while only 27% were taken up by the control liver. A
sevenfold increase was also seen in the adrenals of the estradiol-
treated rats.

Table VI. Comparison of Uptake of ³H-Cholesteryl Linoleyl Ether
           by Male and Female Rats after Injection of Labeled LDL

6 female and 4 male rats in each age group were injected with 60 μg
LDL protein labeled with ³H-cholesteryl linoleyl ether and were
killed 24 h after injection.

Sex	Age, months	Liver	Adrenals	Gonads
		% Dose/organ		
Females	2	29.2	0.35	0.66
Males	2	25.4	0.25	0.60
Females	6	36.3	0.38	0.59
Males	6	35.3	0.40	0.75

Table VII. Effect of 17α-Ethinyl Estradiol on the Hepatic Uptake
            of Human LDL Labeled with ³H-Cholesteryl Linoleyl
            Ether in Male Rats

Male rats 180-200 g of body weight were injected subcutaneously
with 1 ml of 17α-ethinyl estradiol dissolved in propylene glycol
for 5 days. Control rats were injected with propylene glycol.
On the sixth day the rats were injected with LDL (100 μg protein)
labeled with ³H-cholesteryl linoleyl ether. The rats were killed
24 h after injection. Values are means ±S.E.

	Liver	Adrenal
	% of injected dose	
Ethinyl estradiol	81.9 ± 5.5	1.54 ± 0.20
Control	27.2 ± 1.9	0.21 ± 0.03

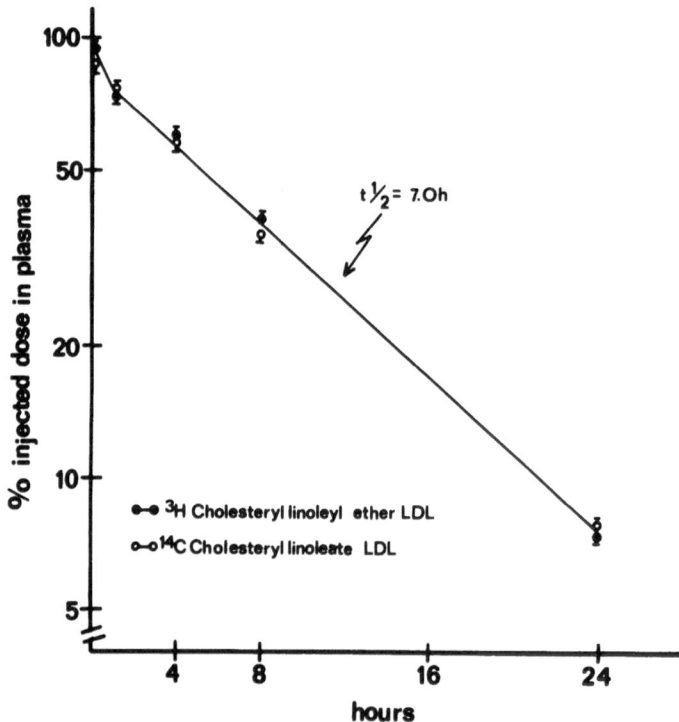

Fig. 4.   Disappearance of LDL label with [³H]cholesteryl linoleyl
          ether and [¹⁴C]cholesteryl linoleate. Values are means
          ±S.E. of 3-11 determinations for each time interval.
          From Stein et al., Biochim.Biophys.Acta, 1981. In press.

DISCUSSION

     In this report we have presented data which validate the pro-
cedure of labeling of lipoproteins with cholesteryl ester and cho-
lesteryl-alkyl-ether. The main advantages of the cholesteryl
linoleyl ether are: 1. chemical purity; 2. high specific radioactivity; 3. nondegradability; 4. recovery in tissue after long
time intervals. In later experiments (manuscript in preparation),
full recovery of the injected label was achieved also up to 2 months.
The advantages of the labeling procedure used are: 1. it utilizes
the physiological pathway for transfer of labeled esterified cholesterol from HDL to VLDL or LDL; 2. the lipoproteins reisolated
after labeling retain their chemical composition and ultrastructural
appearance; 3. their half life in the circulation is not shorter
than that of endogenously labeled lipoprotein. Following injection

into rats, which lack the cholesteryl ester transfer protein, all
the injected label was recovered in the d < 1.019 fraction of
plasma. The prominent role of the rat liver in the removal of VLDL
remnants has been reported previously (2, 13). The highest uptake
of VLDL cholesteryl ester reported so far was 71% of the injected
dose (2); in our study it reached up to 90%. These experiments
point to the almost exclusive removal of VLDL remnants by rat liver.
The findings of the radioautographic reaction associated with the
cytoplasm of parenchymal cells supports our previous findings with
125I-VLDL (13) and provides additional evidence that the VLDL
labeled with cholesteryl linoleyl ether was metabolized in a physiological way. The relatively high recovery of the labeled cholesteryl
ether in the adrenal lends also additional support that the labeled
VLDL has retained its biological properties. Likewise, LDL labeled
by biotransfer with cholesteryl linoleate and its analog cholesteryl
linoleyl ether proved to be a valuable tool to trace the fate of the
nonpolar lipids of LDL. The labeling procedure did not alter the
physiological behavior of the lipoprotein as indicated by its disappearance rate from the circulation. Both ¹⁴C-cholesteryl linoleate
and ³H-cholesteryl linoleyl ether were cleared at the same rate and
the label remaining in the circulation had not been transferred to
HDL. The metabolic fate of the cholesteryl linoleyl ether injected
in the form of LDL differed markedly from that derived from VLDL.
The liver took up only about a third of the injected material and
peripheral tissues (muscle, fat, skin) accounted for the uptake of
almost the same amount as the liver. The tissue distribution of
cholesteryl linoleyl ether labeled LDL in the rat agrees well with
that of ¹⁴C-sucrose labeled LDL in swine (15). Pharmacological
doses of ethinyl estradiol were shown to decrease plasma triglyceride-
rich lipoproteins in the rat (16). This was shown to be due to increased catabolism of these lipoproteins by the liver owing to a
marked increase of receptors that do recognize the B and E apoproteins (17). In a more recent presentation, these authors have
also shown that the hepatocytes account for 90% of the 125I-LDL
taken up by the liver (18). Our present findings with cholesteryl
linoleyl ether labeled LDL indicate that the core lipid follows the
B apoprotein, and that under extreme stimulation of the B receptor
in the liver this organ becomes the main site of LDL catabolism.

ACKNOWLEDGEMENTS

    The excellent technical help of Mrs.Y.Dabach, Mrs.A.Mendeles,
Mrs.M.Ben-Naim and Mr.G.Hollander is gratefully acknowledged.

REFERENCES

1. R.J.Havel, J.C.Goldstein, and M.S.Brown, Lipoprotein and
   lipid transport, in: "Metabolic Control and Disease,"
   P.K.Bondy and L.E.Rosenberg, eds., 8th ed., W.B.Saunders,
   Philadelphia, 393-494 (1980).

2. O.Faergeman and R.J.Havel, Metabolism of cholesteryl esters
   of rat very low density lipoproteins, J.Clin.Invest. 55:
   1210 (1975).

3. M.Krieger, M.S.Brown, J.R.Faust, and J.L.Goldstein, Replacement
   of endogenous cholesteryl esters of low density lipoprotein
   with exogenous cholesteryl linoleate. Reconstitution of a
   biologically active lipoprotein particle, J.Biol.Chem. 253:
   4093 (1978).

4. O.Stein, G.Halperin, and Y.Stein, Biological labeling of very
   low density lipoproteins with cholesteryl linoleyl ether
   and its fate in the intact rat, Biochim.Biophys.Acta 620:
   247 (1980).

5. O.Stein, Y.Stein, A.Fidge, and D.S.Goodman, The metabolism of
   chylomicron cholesteryl ester in rat liver. A combined
   radioautographic-electron microscopic and biochemical
   study, J. Cell Biol. 43: 410 (1969).

6. G.Halperin and S.Gatt, The synthesis of cholesteryl alkyl
   ethers, Steroids 35: 39 (1980).

7. W.Stoll, Eine neue Darstellungsweise von Cholesterinäthern,
   Z.Phys.Chem. 207: 147 (1932).

8. G.Halperin, O.Stein, and Y.Stein, Biological stability of ^{3}H-
   cholesteryl oleyl ether in cultured fibroblasts and intact
   rat, FEBS Letts. 111: 104 (1980).

9. Y.Stein, V.Ebin, H.Bar-On, and O.Stein, Chloroquine induced
   interference with degradation of serum lipoproteins in rat
   liver, studied in vivo and in vitro, Biochim.Biophys.Acta
   486: 286 (1977).

10. N.M.Pattnaik and D.B.Zilversmit, Interaction of cholesteryl
    ester exchange protein with human plasma lipoproteins and
    phospholipid vesicles, J.Biol.Chem. 254: 2782 (1979).

11. P.J.Barter and J.I.Lally, In vitro exchanges of esterified
    cholesterol between serum lipoprotein fractions: Studies
    of humans and rabbits, Metabolism 28: 230 (1979).

12. T.Chajek, and C.J.Fielding, Isolation and characterization of
    a human serum cholesteryl ester transfer protein, Proc.
    Natl.Acad.Sci. USA 75: 3445 (1978).

13. O.Stein, D.Rachmilewitz, L.Sanger, S.Eisenberg, and Y.Stein,
    Metabolism of iodinated very low density lipoprotein in
    the rat; Autoradiographic localization in the liver,
    Biochim.Biophys.Acta 360: 205 (1974).

14. Y.Stein, G.Halperin, and O.Stein, The fate of cholesteryl
    linoleyl ether and cholesteryl linoleate in the intact rat
    after injection of biologically labeled human low density
    lipoprotein, Biochim.Biophys.Acta. In press (1981).

15. R.C.Pittman, A.D.Attie, T.E.Carew, and D.Steinberg, Tissue
    sites of degradation of low density lipoprotein: Application
    of a method for determining the fate of plasma proteins,
    Proc. Natl. Acad. Sci. USA 76: 5345 (1979).
16. R.A.Davis and P.S.Roheim, Pharmacologically induced hypolipid-
    emia. The ethinyl estradiol-treated rat, Atherosclerosis
    30: 293 (1978).
17. Y-s Chao, E.E.Windler, G.Chi Chen, and R.J.Havel, Hepatic
    catabolism of rat and human lipoproteins in rats treated
    with 17α-ethinyl estradiol, J. Biol.Chem. 254: 11360
    (1979).
18. Y-s Chao, A.L.Jones, G.R.Hradek, E.E.Windler, J.S.Mooney and
    R.J.Havel, Electron microscopic morphometric and quantita-
    tive human low density lipoprotein by rat hepatocytes.
    Proc.31st Ann. Meeting Amer. Assoc. Study of Liver Dis.,
    November 1980, Abstr. 10-B, p.10.

INTERACTIONS BETWEEN LIPOPROTEINS AND GLYCOSAMINOGLYCANS

A.L. Catapano, E. Trezzi and R. Fumagalli

Institute of Pharmacology and Pharmacognosy, University of Milan, Via A. Del Sarto 21, Milan, Italy

INTRODUCTION

Heparin and other glycosaminoglycans (GAG) are widely distributed in animal species and humans and may play a relevant role in a number of physiological processes. Heparin and GAG, when intravenously injected, induce a deep, transient modification of the blood clotting system, as well as of the plasma lipids disposition. The effect on the coagulation system is mainly due to the interaction of Heparin with antithrombin III, while that on plasma lipoproteins is usually related to the release of lipoprotein lipases from their binding sites. Plasma lipoproteins, however, interact _in vivo_ and _in vitro_ with GAG and Heparin (1). We shall briefly discuss here the lipoprotein-GAG interactions as a relevant phenomenon in the lipoprotein catabolism as well as a tool in the separation of lipoprotein subclasses.

Interaction between lipoproteins and GAG

Most of the data on the interaction between glycosaminoglycans and lipoproteins have been obtained _in vitro_. Iverius studied the interaction among lipoproteins (VLDL, LDL, HDL) and different GAG covalently bound to sepharose (2). Under controlled conditions (pH, ionic strength) Very Low Density and Low Density Lipoproteins (VLDL and LDL) interacted with GAG and Heparin while High Density Lipoproteins (HDL) did not. High ionic strength effectively inhibited such binding suggesting the presence of ionic interactions (Table 1). Also the acetylation of LDL and VLDL fully suppressed the binding indicating that free amino groups, mainly lysine, may be involved in this interaction. It should be kept in mind, however, that chemical modification may induce conformational changes of the protein therefore data on the chemical modifications should be taken with caution.

Among different GAG Heparin was the most effective in binding the lipoproteins followed by Dermatan sulphate, Heparan sulphate and Chondroitin sulphate, moreover at equal charge density GAG containing L-Iduronic acid interacted more strongly with lipoproteins. Nakashima et al. (3) studied the interaction of GAG separated from human plasma. "Bound" and "Free" GAG precipitated in vitro LDL. Bound GAG also modifyed the hydrocarbon region of pyrene labelled LDL and HDL, not VLDL. These modifications were counteracted by the addition of "free" GAG. This finding suggests, therefore, a role for plasma GAG in mantaining the rheological properties of lipoproteins (increased volume and/or increased volume and/or increased viscosity of the hydrocarbon region). More recently Bihari-Varga et al. showed that Heparin and other GAG, while interacting with, but not precipitating, LDL, moved the phase transition of LDL (melting of the cholesteryl esters core) to higher temperatures (4). The formation of such complexes may be important in determining the catabolism of lipoproteins, for instance increasing the removal rate of lipoproteins by the liver.

GAG may also play a relevant role in trapping lipoproteins at the endothelial level where the triglyceride rich lipoproteins are catabolized by the lipoprotein lipase (1) and in the transport of lipoproteins across the endothelium to the subendothelial space. Anionic probes are, taken up by endothelial cells more readily than catonic probes and most of the plasma proteins have a negative charge. Data obtained by electron microscopy indicate that the transport of these macromolecules should mainly take place through the plasmalemmal vescicles of the endothelial cells (5). The net charge at the level of endothelial cells is, however, not randomly distributed. Anionic sites are, in fact, concentrated at the level of fenestral diaphragms: Heparan sulphate is probably responsible of this net charge (5).

Some evidence has been also obtained that Heparin stimulates the uptake of chylomicrons remnants by parenchymal cells at concentrations of 1 unit per ml in vitro (6). The uptake of HDL_2 is also increased in the same system by Heparin (7). This effect may

TABLE 1

Binding of Lipoproteins to Heparin (dependence upon ionic strength).

	Ionic Strength			
	0.1	0.2	0.5	1.0
VLDL	72	85	8	0
LDL	98	98	5	0
HDL	0	0	0	0

Data from reference 2
Numbers represent the percent of lipoprotein protein bound.

TABLE 2

Chemical composition of VLDL subfractions obtained by Heparin sepharose affinity chromatography.

	Unbound	Bound
Phospholipids	19.2	21.4
Cholesterol	4.6	7.6
Esterified Cholesterol	4.2	9.8
Triglycerides	72.8	61.8
Apo E	+/-	+++

From reference 8

result from either a direct interaction with lipoproteins or an effect on the hepatocyte membranes: in contrast lipoproteins can be released from cellular receptors on cells cultivated in vitro by adequate amounts of GAG. The concentrations required to elicit this effect are, however, of about two orders of magnitude higher than those obtained by heparin injections in vivo (1).

Separation of lipoproteins subclasses by Heparin-Sepharose chromatography

More recently the interaction GAG-lipoproteins has been applied to the separation of lipoprotein subfractions. Shelbourne and Quartford (8) subfractionated VLDL by Heparin-Sepharose affinity chromatography into two subfractions having different chemical composition and apoprotein pattern (Table 2). In particular the unbound fraction contained little, if any, apo E while the bound fraction contained the bulk of apo E, suggesting the presence within VLDL of fractions having different properties. Apo E is in fact responsible for the interaction with Heparin as demonstrated by chemical modification of arginyl residues, and has as well a very high affinity for cellular receptors to lipoproteins.

Weisgraber and Mahley (9) isolated four subfractions of HDL_2, by Heparin sepharose affinity chromatography. The subfractions differed in chemical composition, apoprotein pattern and ability to bind to the LDL receptor in vitro (Table 3). Two fractions contained only apo B, one apo E apo A-I apo A-II and other HDL apoproteins and the unbound fraction, the most abundant, only A-I and A-II and C proteins. Marcel et al. (10) have applied the same system to isolate two subfractions having different ability to act as substrate for the enzyme lecithin cholesterol acyl transferase.

We have also separated four VLDL subfractions by Heparin Sepharose affinity chromatography in presence of divalent ions. Four fractions are obtained that differ in chemical composition and apoprotein pattern (Table 4). Modification of argynil residues with 1-2 cycloexadione drastically modifies the elution pattern (i.e. 99% of VLDL is eluted in fraction 1 and 2).

The interaction between lipoproteins and GAG has been also applied to the separation of GAG and Heparin. Frasson et al. (11) were able to subfractionate Heparin and Heparin sulfate on the basis of their affinity for LDL bound to sepharose. Of interest

TABLE 3

Characteristics of HDL_2 subfractions obtained by
heparin sepharose affinity chromatography.

	Fraction			
	1	2	3	4
Phospholipids	32.8	35.1	24.0	
Total cholesterol	18.4	25.8	33.9	
Triglycerides	3.3	1.3	5.6	
Protein	4.5	36.8	36.5	
Principal apoproteins	A-I,A-II	E,A-I, A-II	B	
Binding to receptors	-	+	+	

From reference 9

TABLE 4

Chemical composition of VLDL subfractions obtained by heparin
sepharose affinity chromatography.

	Fraction			
	1	2	3	4
Phospholipids	11.6	15.2	17.0	17.4
Total cholesterol	9.3	14.1	19.3	15.9
Triglycerides	75.7	66.2	52.7	52.8
Protein	4.5	6.5	10.5	13.2
Apo E	+/-	+	++	+++

the observation that some preparations of Heparin sulphate where bound more tightly than Heparin to LDL. The authors also suggested, on the basis of chemical modifications, that blocks containing L-iduronic acid are more likely responsible for the GAG-LDL interaction.

In similar experiments (but in presence of divalent ions) we have been able to separate Heparin (from pig intestinal mucosa) into two discrete subfractions. The unbound fraction contains both electrophoretically "slow" and "fast" moving Heparin while the bound fraction contains only slow moving Heparin. The physiological significance of this finding is not clear, however, it may be related to the ability of different preparations of Heparin to precipitate lipoproteins containing apo B and E.

SUMMARY AND CONCLUSIONS

The interaction between GAG and lipoproteins is complex and may play a relevant role in regulating the catabolism of plasma lipoproteins. The interaction appears to be mediated by electrostatic forces and is specific for the protein moiety (apo E and B in particular), in vivo it may control several processes:
a) trapping of lipoproteins at the endothelial surface
b) catabolism of lipoproteins via a specific receptor
c) transport of lipoprotein across the endothelium
d) physical state of lipoproteins in the blood stream.

In addition the interaction GAG-lipoproteins offers a tool in subfractionating classes of lipoproteins that have different chemical composition and apoprotein pattern, and may also differ in origin and/or catabolism.

ACKNOWLEDGEMENTS

The skillful assistance of Mrs. Aurora Maccalli Rossoni in typing the manuscript is acknowledged.

This research was supported in part by a grant from C.N.R. No. 79.01086.83

REFERENCES

1. G.C. Ghiselli and A.L. Catapano, Glycosaminoglycans and lipoprotein metabolism: an overview, Pharmacol. Res. Comm. 11:571 (1979).

2. P.H. Iverius, The interaction between human plasma lipopro-
    teins and connective tissue glycosaminoglycans, J. Biol.
    Chem. 247:2607 (1972).

3. Y. Nakashima, N. Di Ferrante, R.L. Jackson, and H.J. Pownall
    The interaction of human plasma glycosaminoglycans with plas-
    ma lipoproteins, J. Biol. Chem. 250:5386 (1975).

4. M. Bihari-Varga, M. Sztatisz, and S. Gal, Changes in the
    physical behavior of low density lipoproteins in the presence
    of glycosaminoglycans and high density lipoproteins, Athero-
    sclerosis 39:19 (1981).

5. M. Simionescu, Structural and functional differentiation of
    microvascular endothelium in Blood cells and vessel walls,
    Ciba Foundation Symposium 71, Elsevier, Amsterdam, pp. 39
    (1980).

6. C.M. Floren and D. Nilsson, Binding interiorizatio and degra-
    dation of cholesteryl ester labelled chylomicrons remnants
    particles by rat hepatocyte monolayers, Bioch. J. 168:483
    (1977).

7. G.C. Ghiselli, R. Angelucci, A. Regazzoni, and C.R. Sirtori,
    Metabolism of HDL_2 and HDL_3 cholesterol by monolayers of rat
    hepatocytes, FEBS Letters 125:60 (1981).

8. F.A. Shelburne and S.M. Quartford, The interaction of heparin
    with and apoprotein of human very low density lipoproteins,
    J. Clin. Invest. 60:944 (1977).

9. K.M. Weisgraber and R.W. Mahley, Subfractionation of human
    high density lipoproteins by heparin sepharose affinity chro-
    matography, J. Lipid Res. 21:316 (1980).

10. Y.L. Marcel, C. Vezina, D. Emond, and S. Ginzaburo, Hetero-
    geneity of human high density lipoproteins: presence of lipo-
    proteins with and without apo E and their roles as substrates
    for lecithin cholesterol acyltransferase reation, Proc. Natl.
    Acad. Sci. USA 77:2964 (1980).

11. L.A. Fransson and B.G. Johansson, Glycosaminoglycan-Lipopro-
    tein interactions: 1. Effect of Uronic acid composition and
    charge density of the glycan, Int. J. Biol. Macromol. 3:25
    (1981).

PLASMA LIPOPROTEIN CHANGES INDUCED BY DIETS AND DRUGS

Cesare R. Sirtori

Center E. Grossi Paoletti and Chemotherapy Chair

University of Milan, 20129 Milan, Italy

INTRODUCTION

Diet and/or drugs are commonly used for the management of hyperlipoproteinemias (1). On the other hand diets and, at times, drugs are also prescribed to normolipidemic individuals, in the hope of modifying risk factors for atherosclerosis. Among these, low high density lipoprotein (HDL) cholesterol levels (2), obesity (3), hypertension (4) and increased platelet aggregation (5) are often managed by diet and drug treatments which may modify plasma lipid transport.

Recent advances in the knowledge of lipoprotein physiology, as also reported in most presentations during this course, have allowed a better understanding of the significance of lipoprotein changes during diet and drug treatments, aimed to preventing and possibly reversing the atherosclerotic process. In this review, findings related to diet and drugs of current use will be analyzed, both in view of their impact on lipoprotein physiology and of their significance in atherosclerosis prevention and regression.

DIETARY COMPONENTS AFFECTING LIPOPROTEIN METABOLISM AND ATHEROSCLEROSIS REGRESSION

Several dietary factors may influence lipoprotein metabolism. Meal distribution, energy content, as well as leisure items

163

(e.g., coffee and ethanol), besides the major nutritional components, may be relevant in the overall changes of atherosclerosis
risk factors.

Meal distribution - Meal frequency has been known for sometime to
affect body weight gain. "Nibbling", i.e. eating small amounts of
food many times during the day, elicits more marked weight gains
in rats, as compared to "gorging", i.e. overeating once in the
day time (6). Human experiments have shown that eating the same
amount of calories distributed into 3-6 meals will maintain
stable plasma glucose and insulin levels, whereas eating them in
one single meal will result in a marked rise of plasma free fatty
acids (FFA) before the meal and of plasma glucose thereafter (7).

Plasma lipid and lipoprotein changes resulting from these
different eating schedules have not been studied in great detail.
Recenty, however, Kay et al. (8) showed that a fat load given as
a single daily meal may result in a higher total cholesterol and
lower HDL cholesterol (HDL-C) and apo AI levels, as compared to
the same fat intake distributed into six meals. In this latter
case, a significant increase of apo AI during the day time may be
observed.

Energy content - Obesity is an established risk factor for atherosclerosis, particularly when associated with hyperlipidemia and/
/or hypertension (3). Weight reduction in obese hypertriglyceridemic individuals leads, in most cases, to triglyceride (TG)
reduction, at times back to normal limits (9).

The occurrence of lipoprotein changes other than reduction
of very low density lipoprotein (VLDL) TG following weight reduction is disputed. Enhanced VLDL catabolism may result in increased LDL cholesterol levels (10), particularly in subjects with
very low levels of these lipoproteins. Changes of HDL-C levels
following energy, and particularly carbohydrate (CHO), reduced
diets have been reported by some Authors and not by others. If
tissue cholesterol is mobilized during the reduction of body mass
(in view of the known direct relationship between total body
cholesterol and body weight) (11), then excess cholesterol is
likely be mobilized via HDL (12). Nestel and Miller (13) showed
increased specific activity of HDL-C in obese individuals undergoing weight reduction. Whether this may also result in increased
absolute levels of HDL-C is controversial. Decreased HDL-C levels
after weight reducing diets were noted by Thompson et al. (14);
no significant change was reported by Hulley et al. (15), where-

as, more recently, Contaldo et al. and Streja et al. (17) showed both decreased plasma TG and increased HDL-C in morbidly obese patients following marked weight reductions. The last Authors underline that a significant HDL-C increase may be found only after sizable weight losses.

Calorie and CHO restrictions for the treatment of hypertriglyceridemia (be it associated with weight reduction or not), generally fail to increase HDL-C levels. In a recent long term study, Witztum et al. (18) showed negligible HDL-C changes after stabilization of type IV hyperlipidemic patients on low calorie, CHO and ethanol diets, effectively reducing plasma TG.

Polyunsaturated fatty acids - The addition of polyunsaturated fatty acids, with a resulting increased polyunsaturated/saturated (P/S) ratio, is common practice in the management of most forms of hyperlipoproteinemia. On the other hand, this dietary approach is believed to exert a protective effect against coronary heart disease (CHD), as shown by longitudinal controlled studies (19,20) and by a recent retrospective survey (21).

Both plasma cholesterol and TGs are reduced after diets with increased P/S ratios. The mechanism of the cholesterol reduction is still unclear, in that different findings have been reported from cholesterol balance studies in normal (22) and type II hyperlipidemic patients (23). In the former, an increased neutral steroid and bile acid excretion was detected, thus supporting the indication of increased lithogenicity following long term treatments (24). In type II patients, in contrast, decreased plasma cholesterol is not associated with any change, either in fecal sterol excretion, or in the slope of the plasma cholesterol specific activity decay curve (23). This latter observation would suggest that some body cholesterol redistribution occurs following the diet. TG reduction, on the other hand, is most likely due to a diminished incorporation of P fatty acids into VLDL TGs, as compared to S fatty acids (25).

Plasma lipoprotein, and particularly HDL-C changes, following treatments with high P/S diets have been the object of several recent investigations. In a detailed study on HDL composition and metabolism in normolipidemic subjects administered an experimental diet with a very high P/S ratio (4.0), Shepherd et al. (26) demonstrated both a striking decrease of total cholesterol and of HDL-C levels. Turnover studies with iodinated apo AI in these subjects failed to disclose differences in catabolism, the decrease of HDL apoproteins in these subjects being seemingly due to reduced synthesis. Other Authors have shown similar changes in hyperlipidemic subjects. Vessby et al. (27) (Table I) again demonstrated decreased HDL-C and apo AI in type II and IV hyperlipidemic patients undergoing an ordinary P/S 2.0 therapeutic diet. The mechanism of the decrease is unclear, as is its clinical significance. An intestinal effect of polyunsaturated fatty acids on the secretion of apo AI has been suggested (28). It may be noted, however, that other studies in normo- and hyperlipidemic patients have either not shown a decrease of HDL-C, or indicated an increase (29,30). A modification of other risk factors for atherosclerosis was attempted in the latter studies.

TABLE I

PLASMA LIPOPROTEIN CHANGES FOLLOWING A HIGH P/S DIET IN HYPERLIPOPROTEINEMIAS[27]

9 pts (3 m. and 6 f., 2 type II A, 2 type II B, 1 type III, and 4 type IV);
isocaloric diet, 35% L, 20% P, 45% C, P/S 2.0

		Pre-diet	mg/dl	Post-diet
Total Chol.		317 $\pm$ 60		237 $\pm$ 28[++]
Total TG		261 $\pm$ 91		187 $\pm$ 66[++]
VLDL Chol.		50.6 $\pm$ 51.3		24.0 $\pm$ 10.0[++]
LDL Chol.		204 $\pm$ 52.5		162 $\pm$ 27.4[++]
HDL Chol.		51.7 $\pm$ 13.5		43.2 $\pm$ 11.2[++]
Apo AI		1.03 $\pm$ 0.21		0.93 $\pm$ 0.20[+]
Apo B	Arb. Units	1.89 $\pm$ 0.59		1.40 $\pm$ 0.50[++]

[+] $p < 0.05$ [++] $p < 0.01$

HDL isolated from subjects administered diets of different fatty acid composition show, on the other hand, a similar effectiveness on cell cholesterol removal (31). These findings indicate that the amount and rate of removal of cellular cholesterol induced by HDL are independent of compositional changes effected by dietary perturbations.

Cholesterol - Cholesterol feeding is still the most common dietary mean for the experimental induction of hyperlipidemia. In spite of the obvious interest in this topic, clinical studies have been relatively limited. In general, humans appear to be definitely more resistent to the effects of dietary cholesterol than animals.

The effects of cholesterol feeding in humans are markedly influenced by the way of administration. When, in fact, cholesterol is given within a liquid formula diet, changes of cholesterolemia are remarkable (32). Formulas calculating the cholesteremic effects of dietary cholesterol have been extrapolated from these studies. The effect is a direct function of the square root of total daily intake in mg (33). On the other hand, when cholesterol is administered within solid food items, effects are far less remarkable. With the exception of scattered reports on extreme cholesterol changes induced by egg rich diets (34), most recent studies on the cholesterol addition to solid foods suggest a negligible activity on cholesterolemia (35,36,37). More recently, Bronsgeest-Schoute et al. (38) have indicated, from prospective and retrospective studies, an individual sensitivity to the hypercholesterolemic effect of dietary cholesterol.

Considerable interest has been, lately, raised by lipoprotein changes occurring after acute or chronic cholesterol administrations to humans. In humans, similarly to rabbits, but in the absence of changes in total cholesterolemia, a cholesterol enriched VLDL fraction may be detected following large cholesterol intakes (39). An even more interesting finding, in the same conditions, is that of a cholesterol enriched HDL subfraction (HDL-I), exerting a powerful inhibitory activity on the cholesterol synthesis of human cultured fibroblasts (40). The formation of increased amounts of HDL-I (or HDL_c) following dietary cholesterol may perhaps provide an endogenous protective mechanism against hypercholesterolemia development.

Undigestible fibers - The use of undigestible fibers in the diet, i.e. plant polysaccharides and lignin, resistant to the hydrolysis by the digestive enzymes (41), has recently gained large acceptance. Epidemiological studies (42) have suggested that a low fiber intake is linked to the development of the atherosclerotic disease and of large bowel disorders in the Western World. Fiber administration, moreover, when in limited amounts, is quite acceptable and free of significant side-effects.

The hypolipidemic activity of dietary fibers is controversial. With the exception of some types of fiber, e.g. pectin (43) and particularly guar gum (44), most studies have not provided a definitive answer on the plasma lipid lowering activity of these dietary components.

Clinical controlled studies on the effects of pectin on body cholesterol balance, provided negative results (45). These data should be, however, considered with caution, since only endogenous sterol excretion was reported, and not total excretion. Animal studies have, in fact, demonstrated a significant absorption of bile acids to the polysaccharide components of undigestible fibers (46). This will, in turn, result in an enhanced fecal elimination of neutral sterols and bile acids. Accelerated regression of diet established atherosclerosis in primates has been described following the addition of the saponin components of fibers to the lipid lowering diet (47).

Severely hypercholesterolemic patients, when treated with large amounts of guar gum, added to the diet under the form of crispbread, show a significant reduction of plasma cholesterol levels (48). The effect is not less than with cholestyramine. On the other hand, some caution has been suggested in the use of this mucillagenous material, since early animal reports indicated significant pathological changes in the gut following guar administration (49).

Controlled prospective studies in healthy individuals given bran or citrus pectin confirm that the former type of fibre is ineffective in lowering cholesterolemia (50). Citrus pectin, on the other hand, has some therapeutic activity on plasma cholesterol as well as on fecal elimination, particularly of bile acids (50). Lipoprotein changes following fiber enriched diets are usually restricted to LDL-cholesterol. Changes in HDL-C, although occasionally reported, have not been generally confirmed.

Proteins - A different activity of proteins of vegetable origin on plasma lipid levels, as compared to animal proteins, was suggested early this century by Ignatowski (51). Later, scattered reports, also indicating a hypocholesterolemic effect of vegetable versus animal proteins (52,53), received little attention, particularly in view of the higher interest in the dietary lipid theory of atherosclerosis.

In more recent years, experimental studies in rabbits (54) as well as clinical studies, mostly carried out by our research group (55,56), have again suggested a specific hypocholesterolemic activity of vegetable proteins in the diet. Most of the experiments have examined the partial or complete substitution of animal proteins (casein in the animal studies) with soybean proteins. In humans, textured products, i.e. commercial delipidated preparations containing re-structured proteins with minor amounts of other components, appear to be more effective than soybean "isolates" (57).

Many different studies have been carried out by us with the soybean protein diet. After an initial comparison with a standard low lipid-low cholesterol diet (55), the effect of the addition of saturated fatty acids to the diet was examined (58). More recently, an eight week study on out-patients, carried out in several Italian and one Swiss Center was reported (59). This study (Fig. 1) provided clear evidence of the effectiveness of the regimen also in these conditions. It also indicated that the diet is selectively effective on LDL cholesterol levels, while minimally modifying TG and VLDL. No effect on HDL-C was seen in our studies.

Contrasting reports on the soybean diet have been recently published (60). Several shortcomings of this last study have been pointed out (61). An interesting difference between the activity of beef and of soybean proteins on HDL-C was, however, observed: the former regimen decreased HDL-C, whereas the latter seemed to reverse the effect.

The mechanism of action of the soybean diet is still obscure. Animal studies have suggested that an increased fecal excretion of neutral steroids and bile acids may occur, as compared to a similar diet with casein (62,63). Similar findings were reported in infants administered a soybean milk (64). On the other hand, cholesterol balance studies carried out by our group failed to disclose any effect of the diet on either fecal steroid excretion or in the slope of the plasma cholesterol specific activity decay curve, in spite of the dramatic plasma cholesterol reduction (65). The explanation is obviously difficult; it should only be noted that somewhat similar data were reported with high P/S diets in hypercholesterolemic patients (23).

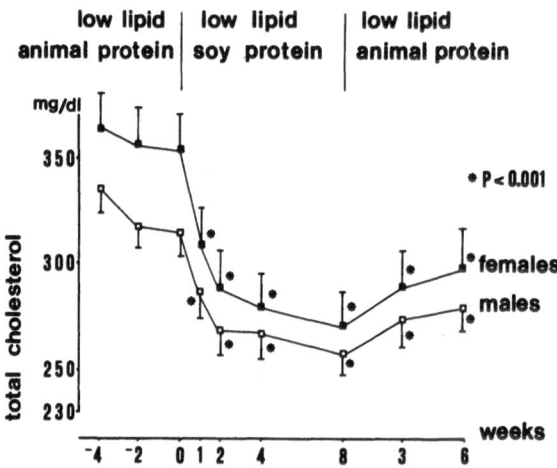

Fig. 1 - Plasma total cholesterol changes in 127 adult patients with type II hyperlipoproteinemia, followed for 4 weeks on a standard low cholesterol-high P/S regimen, 8 weeks on the soybean protein diet, and 6 more weeks on the standard regimen (59).

Ethanol - The effects of ethanol on lipoprotein metabolism and on the development of the atherosclerotic disease, have received considerable attention in the past few years. Ethanol has been generally considered as a hypertriglyceridemic factor, both in man and experimental animals (66,67) particularly when added to lipid rich regimens (68).

More recently, epidemiological studies on the plasma lipoprotein profile following different nutrients have indicated that the dietary ethanol content bears a straight relationship with plasma HDL-C levels (69). In view of the putative protective effect of raised HDL-C levels in atherosclerosis prevention (70), retrospective longitudinal studies have examined the correlation between ethanol intake and incidence and severity of CHD. In Hawaii, studies of Japanese-American citizens have shown a negative correlation between daily ethanol intake and incidence of myocardial infarction, not of angina pectoris (71). In another retrospective study, pooling middle aged male subjects from European and non European nations, a significant negative correlation was observed between yearly red wine consumption and incidence of CHD (72). This report, somewhat anecdotal (Fig. 2), also suggests that the consumption of spirits other than red wine does not have the same protective effect. The presence of some "protective" factor in the diet of Latin European Countries has been supported by the recent results of a study on coronary incidence in France (73). In spite of a similar prevalence of acknowledged CHD risk factors, in fact, the incidence of the disease in this country is less than half that found elsewhere, e.g. in the US or Scandinavia.

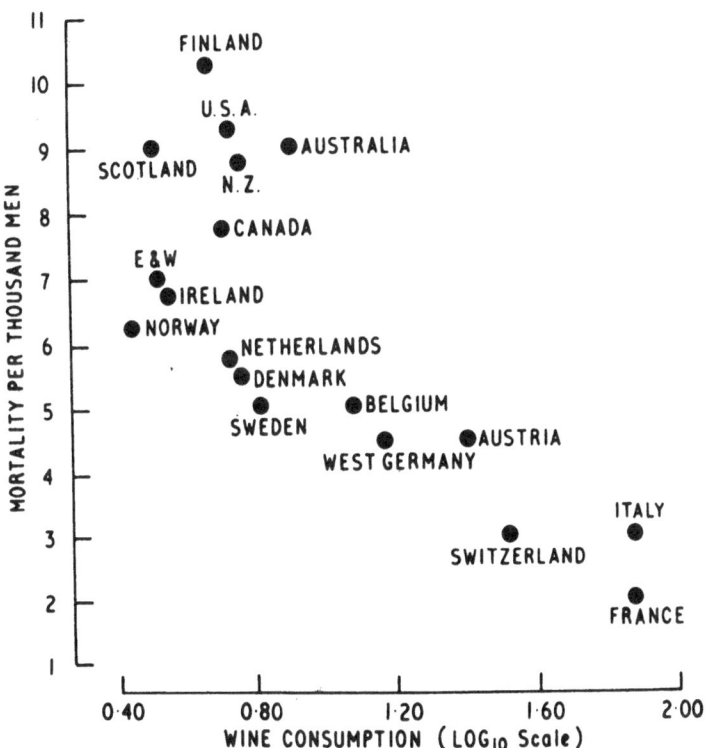

Fig. 2 - Correlation between red wine consumption ($\log_{10}$ of liters/year) and cardiovascular mortality in Western countries (72).

The mechanisms underlying raised HDL-C levels following ethanol have been extensively analyzed. Lipoprotein lipase (LPL)
activities, both in the adipose tissue and in muscles, are
increased following ethanol (74). In this way, both an increased
TG synthesis and an accelerated VLDL catabolism are induced by
ethanol. VLDL surface components are transferred to the HDL
density range (75), thus also raising the HDL_2/HDL_3 ratio (74).
These effects occur at relatively low dietary ethanol intakes
(e.g., around 30 ml/day). Larger intakes should be discouraged,
in view of the known negative effects on blood pressure and liver
function.

Coffee - An association between consumption of coffee and incidence of atherosclerotic disease has never been clearly confirmed
by epidemiological studies (76). On the other hand, recent clinical studies underline significant haemodynamic effects of coffee
intake, particularly in non-habituated subjects (77). In some
studies, coffee intake and cigarette smoke, when associated, have
been shown to exert a positive interaction on the incidence of
atherosclerosis (78). Experimental studies have not been contribu-
tory to the understanding of a possible role for coffee or tea in
atherosclerosis development. Tea is inactive in a hypercholesterolemic rat model (79). On the other hand, when rhesus monkeys,
fed an atherogenic diet, are given coffee as 50% of their fluid
intake, no significant change in plasma lipoproteins and/or atherosclerosis progression is observed (80).

In the only detailed study on the human plasma lipoprotein
profile, retrospective data were collected in subjects known to
consume large (> 5 cups per day) or small (< 5 cups per day) a-
mounts of coffee (81). Whereas a large coffee intake seems to
have a slight positive effect on HDL-C and a negative influence
on LDL-C levels, concomitant smoking inverts the pattern. Heavy
cigarette smokers and coffee drinkers have, in fact, the highest
LDL-C and the lowest HDL-C among the studied individuals.

DRUGS AFFECTING LIPOPROTEIN METABOLISM AND ATHEROSCLEROSIS REGRES-
SION

Studies on the effects of drugs on lipoprotein metabolism
and/or atherosclerosis progression have been numerous, particularly in the case of hypolipidemic agents. On the other hand, considerable interest has been recently devoted to the metabolic effects of drugs chronically used in atherosclerosis prevention

(hypotensive agents, drugs affecting platelets, etc.) as well as of environmental contaminants. Particular emphasis will be given to studies analyzing the mechanism of action of hypolipidemic agents. Available information about the effects of drugs, chronically administered for other indications, and about environmental contaminants will be also provided.

Absorbable hypolipidemic drugs

Clofibrate and the "catabolic agents" - Clofibrate (CPIB) is still the most widely used hypolipidemic agent in the Western World. The mode of action of CPIB, developed over 20 years ago, has remained surprisingly obscure up to recent years. With the improved knowledge on the mechanism, the possibility of side-effects occuring after prolonged treatments has emerged (82). These recent reports have contributed to raising significant doubts on the long term efficacy and safety of the compound (83).

CPIB and related compounds (bezafibrate, procetofene, gemfibrozil) should be classified as "catabolic" drugs, in view of their significant stimulatory activity on lipoprotein catabolism. All of them, in fact, even when given for a short time, considerably increase the catabolism of VLDL, by stimulating LPL activity (84). In some cases, i.e., bezafibrate, also the hepatic lipase activity may be raised (85). The enhanced TG breakdown results in hypotriglyceridemia and may also be followed by raised LDL-C and HDL-C levels (86). This last change may, in turn, enhance cholesterol mobilization from tissues, as shown in experimental animals (87) and in man (88).

The "catabolic" activity of CPIB and of related compounds is probably responsible for most of the reported side-effects and, possibly, also for the long-term toxicity shown in rodents. Cholesterol mobilization from tissues, probably by way of HDL (12) may lead, in fact, to bile supersaturation, with enhanced lithogenicity. Although other hypolipidemic agents seem to share this property of CPIB to a lesser extent (89), more conclusive evidence on the relative risk of lithogenicity should be obtained by studies of fecal cholesterol excretion (88).

Plasma lipoprotein changes occurring after treatment may also be ascribed to the stimulated LPL activity. In subjects with hypertriglyceridemia and low LDL-C levels, an increase of LDL-C may be observed following treatment (86). As above pointed out, also the HDL rise, not seen in all studies (90), is a likely consequence of activated tissue lipolysis; both HDL_2 and HDL_3 may increase after treatment (91).

Aside from the predictable lithogenicity, the recently published long-term studies with CPIB in the prevention of CHD (82) have also given considerable worry on other major side-effects, e.g. raised incidence of tumors. In parallel to the clinical studies, not in agreement in this respect with previous American data on secondary coronary prevention (92), rodent experiments have shown that CPIB and related compounds, all inducers of peroxisome proliferation in the liver (93), may cause liver tumors following long-term treatments (94).

This last toxic effect, apparently only restricted to rodents, is possibly also related to the "catabolic" mode of action of CPIB and related compounds. The stimulation of TG breakdown may, in fact, cause an overload of fatty acids in the liver. During oxidation of these, H_2O_2 is produced at several steps of the β-oxidative chain as shown in Fig. 3. In the highly sensitive rodent liver, this will lead to peroxisome proliferation and, following long-term treatment, to tumor development. In humans, differently from rodents, only mitochondria, with a raised number of cristae, are proliferated after CPIB (95). This hypothetical unifying mechanism of CPIB and related compounds, explaining lithogenicity, lipoprotein changes and rodent tumors all by one mechanism, i.e. increased LPL activity, is worth more detailed analysis. Other treatments, e.g. administrations of large amounts of fat or of aspirin to rodents, also lead to peroxisome proliferation (96,97). In the first case, the biochemical mechanism is clear; in the second, the TG lowering activity of aspirin in these animals should be remembered (97). Some exceptions have been, however, presented. The p-fluoro derivative of CPIB, although hypolipidemic, apparently does not proliferate peroxisomes (98) (Fig. 4).

Nicotinic acid - The mode of action of nicotinic acid (NA) and derivatives is somewhat different from that of CPIB. Both compounds activate, in fact, LPL (99), but NA exerts a more pronounced inhibitory activity on the adipose tissue hormone sensitive lipase (100). On the other hand, comparative studies of CPIB and NA on plasma lipoproteins in type IV subjects, provided quite similar results, with a positive interaction of the two agents (86). NA seems to be more effective on plasma cholesterol levels, possibly by enhancing catabolism of LDL (101).

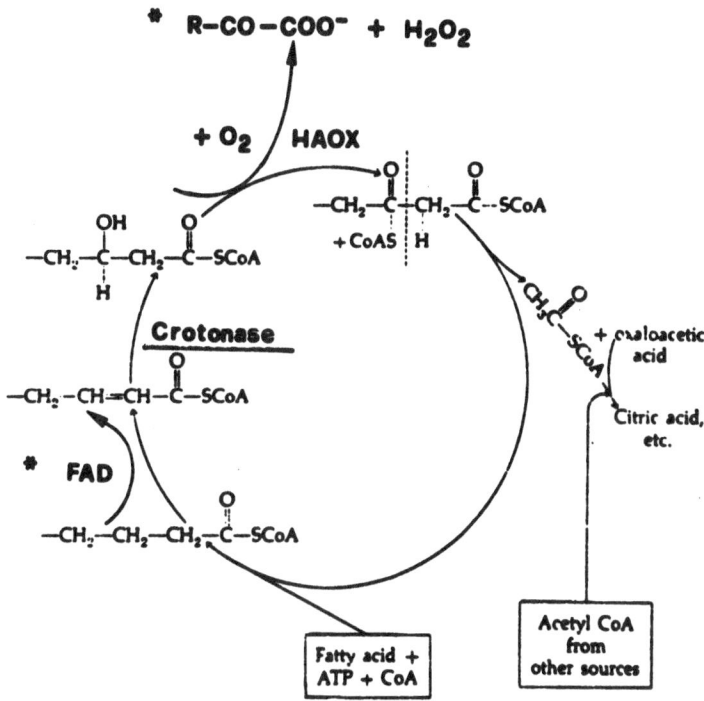

Fig. 3 - Proposed mechanism of H_2O_2 production following hypo-lipidemic drugs treatments. * indicate metabolic sites of H_2O_2 production. Crotonase is tentatively identified with the 80,000 MW protein observed in liver microsomes of drug treated animals. (Modified from "Principles of Biochemistry", A. White et al. eds, 6th Ed., Mc Graw-Hill, New York, 1978).

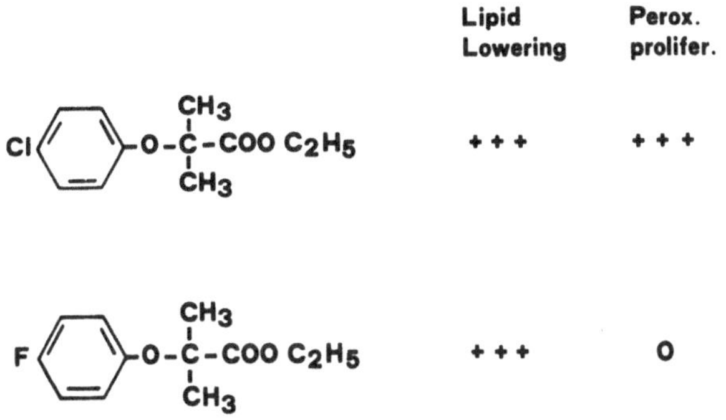

Fig. 4 - Difference in peroxisome proliferating properties between clofibrate and its fluoro-analog.

Contrasting results have been provided on the activity of NA and derivatives in normolipidemic and hyperlipidemic individuals. Whereas, in fact, in normolipidemic volunteers a marked rise of HDL-C levels, with an inversion of the HDL_2/HDL_3 ratio has been described (102), HDL changes are far less significant in hyperlipidemic patients (103).

Since the activity of NA on lipoprotein catabolism is probably similar to that of CPIB, comparative studies have been carried out with the two agents on fecal steroid elimination. It was reported, by some Authors (104) that NA induces a lesser degree of cholesterol saturation in bile, thus giving rise to a lower lithogenic risk. In contrast, Angelin et al. (105), by steroid balance studies, recently indicated that NA, similarly to CPIB, increases lithogenicity. The results of the Coronary Drug Project study have shown that NA (at relatively low doses) may be less lithogenic than CPIB (92).

Tiadenol - This new hypolipidemic agent, bis-(hydroxyethyl 2-ethylthio) 1-10 decane , chemically unrelated to CPIB, has exerted a significant therapeutic activity both in type II and IV hyperlipidemias (106,107). Administration to rodents, similarly to CPIB, induces liver enlargement and peroxisome proliferation (108). On the other hand, differently from CPIB and related compounds, tiadenol (T) does not induce cholesterol mobilization, but rather causes tissue cholesterol retention (109).

Clinical studies with T in type IV patients suggest that, differently from CPIB and NA, T may not stimulate LPL activity. Studies on the composition and apoprotein structure of VLDL show a remarkable decrease of both cholesterol and TG content, together with a significant reduction of apoprotein E (110). These changes may indicate a reduced liver secretion of TG rich lipoproteins. Apo E is, in fact, considered the only apoprotein exclusively secreted by the liver (111). In the same study, T was effective in the sucrose induced hypertriglyceridemia of diet sensitive patients. No changes of the LPL activity were detected in these patients after treatment with T (110).

Differently from CPIB and the "catabolic" drugs, T may not affect lipoprotein catabolism, but rather reduce lipoprotein, and particularly VLDL, secretion from the liver. Studies on the possible lithogenicity of the compound are of particular interest, in view of the hypothetical mechanisms underlying this side effect of other hypolipidemic drugs.

Metformin - Considerable attention has been recently paid to this anti-diabetic agent (N,N-dimethyl biguanide), in view of its capacity to reduce atheromatosis in cholesterol fed rabbits, without significantly modifying plasma lipid levels (112). Apolipoprotein changes, particularly with reduced apo E in VLDL and increased apo AI in VLDL and HDL, have been described (113). Metformin (M), similarly to phenformin (114), does not activate LPL.

Investigations on the mode of action of M have considered that this drug, similarly to other biguanides, is selectively localized in the intestinal wall (116). The apoprotein composition of lymph in cholesterol fed rabbits treated with metformin shows marked changes in the same direction as in plasma, i.e. decreased apoprotein E and increased apo AI (117).

Apoprotein modifications similar to those observed in cholesterol-fed rabbits are also seen in man. In type III patients, a reduced apoprotein E in VLDL is observed after treatment (113). Moreover, both in hyperlipidemic and normolipidemic individuals, increased HDL-C levels and raised AI/AII ratios in HDL are observed (118). Since similar changes are noted after anion exchange resin treatment (see later), these findings suggest that drugs acting on intestinal lipoprotein biosynthesis may modify the apoprotein composition of HDL.

The interest in a possible anti-atherosclerotic effect of metformin, independent of plasma lipid lowering, is considerable. Clinical, not controlled studies, have suggested remarkable improvements, particularly in the peripheral circulation, following M treatment (119). Moreover, a reduced platelet aggregability both in hyperlipidemic animals and in humans has been described (120). Clinical controlled trials are being carried out, in order to assess the potential anti-atherosclerotic effect of metformin in patients with angiographically proven peripheral vascular disease.

Anion Exchange Resins - Synthetic compounds covalently binding bile acids in the duodenum are an effective mode of treatment of hypercholesterolemias (121). In controlled studies of post-myocardial infarction patients with hypercholesterolemia, colestipol treatment significantly reduced cardiovascular mortality (122). Moreover, in numerous animal investigations, cholestyramine has accelerated the dietary regression of atherosclerosis (123); more recently, a "protective" effect has been proven also in cases where the dietary induction was maintained (124).

The mechanism of action of anion exchange resins appears quite simple at first consideration. Bile acids, when bound, are not reabsorbed and are lost in the feces. In this way, after prolonged treatment (but a significant plasma cholesterol lowering effect may be noted already after one week), the circulating sterol pool is reduced, with consequent hypocholesterolemia.

On the other hand, this simple mechanism of action does not provide a satisfactory explanation to some questions raised by studies on anion exchange resins. In the first place, it does not clarify the observed insensitivity of homozygous type IIa patients (125). In spite of the significant increase of fecal bile acid loss, the hypocholesterolemic activity is practically absent in these patients. Moreover, in all cases, the intestinal loss of bile acids is followed by an immediate rise of liver cholesterol biosynthesis and 7-α-hydroxylation (126); these biochemical changes, although effectively antagonizing the primary drug effect, still fail to counteract the hypocholesterolemic activity.

In view of these observations, other studies have been carried out to better define the lipoprotein changes following anion exchange resin treatment. An increased VLDL biosynthesis has been described by several Authors transiently or permanently raising plasma TG levels after treatment (127). These changes are apparently consequent to the bile acid loss; similar findings are, in fact, reported in monkeys following bile diversion and are corrected by the re-infusion of bile acids (128). Bile acid flux may be negatively correlated with VLDL biosynthesis; indirect evidence for this hypothesis is provided by the hypotriglyceridemic activity of bile acid administration (129).

The increased VLDL biosynthesis following anion exchange resin treatment may markedly modify low density lipoprotein turnover. LDL apoproteins are, in fact, derived from two sources, the major fraction in normal subjects coming from VLDL breakdown (130), the rest from an independent input (131). In type II hyperlipoproteinemia, a significantly larger fraction derives from the independent input (131). Animal studies, particularly in rabbits, have suggested that independently synthesized LDL may have a lower turnover rate as compared to the fraction derived from VLDL (132) (Fig. 5). Although the possible clinical significance of these findings is yet to be established, studies in type II patients treated with cholestyramine showed a higher rate of LDL catabolism following treatment (133). Moreover, Shepherd et al. (134) recently demonstrated that a larger fraction of LDL in cholestyramine treated hypercholesterolemic patients is catabolized by a receptor-dependent pathway (Table II). These findings, also showing some compositional changes of LDL following treatment, are consistent with a modification of the LDL catabolic pattern following cholestyramine. The receptor-dependent pathway, normally accounting for the catabolism of 33% of LDL (134), is reduced in hypercholesterolemic individuals and increased by anion exchange resin treatment. In this way, the amount of LDL

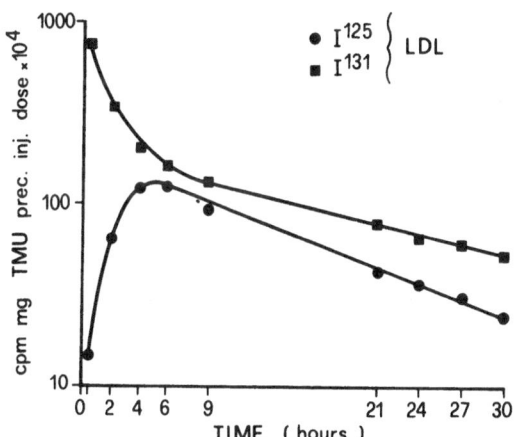

Fig. 5 - Apo B SA time curves in the LDL density range after simultaneous ^{125}I-VLDL$_{II}$ (d < 1.1019) and ^{131}I-LDL (d 1.019-1.063) injection into normal fasted rabbits. The reported representative experiment indicates that the two curves do not meet at any time interval, thus suggesting heterogeneity of native LDL, which is comprehensive of a fraction originating from VLDL and of a more slowly turning over fraction, probably of independent biosynthesis (132).

TABLE II

INCREASE OF RECEPTOR MEDIATED LDL CATABOLISM FOLLOWING
CHOLESTYRAMINE TREATMENT[134]

5 type II female patients, 24 g/day of cholestyramine for 4 weeks			

PLASMA LIPIDS	Start		End
		mg/dl	
Plasma Chol. Total	432	$\pm$ 81	313 $\pm$ 45[++]
LDL	347	$\pm$ 81	232 $\pm$ 50[++]
Apo LDL	228	$\pm$ 83	178 $\pm$ 62[+]

APO LDL CATABOLISM

Fractional catabolic rate   pools/day

Total	0.185 ± 0.046	0.239 ± 0.044[++]
receptor mediated	0.033 ± 0.011	0.067 ± 0.004[++]

Absolute catabolic rate   mg/day

Total	16.1 ± 3.0	16.5 ± 4.3
receptor mediated	2.8 ± 1.1	4.8 ± 1.8

Receptor independent catabolism   mg/kg/day

of LDL cholesterol	20.5 ± 4.8	15.4 ± 3.3[++]

[+] $p < 0.05$ and [++] $p < 0.01$ versus pre-treatment

disposed of by the scavenger pathway, i.e. tissue macrophages (including arterial wall macrophages) may be reduced.

The correlation between increased VLDL biosynthesis and accelerated LDL catabolism still needs to be proven. Different hypotheses, i.e. an increased affinity of LDL for hepatic receptors following drug treatment, have also been suggested (135). Other apoprotein modifications, i.e. raised AI/AII ratio in HDL may also be detected after treatment (136). In any case, a simple intestinal effect of anion exchange resins probably fails to explain the complex mode of action of these highly effective medications for atherosclerosis regression.

Hypotensive agents - Recent interest has been devoted to the activity of chronically used drugs on plasma lipoproteins. Both beta-blockers and diuretics seem to share the common property of raising VLDL TG and of lowering HDL-C (137-39). These effects, especially seen with non-selective beta-blockers (e.g., propranolol), may be particularly significant in the case of mildly hypertriglyceridemic individuals (140). The mechanism is not clear: an altered control of tissue lipolysis may be a factor; it has also been suggested that a reduction of LPL activity may be effected by beta-blocking drugs (141). Previous studies from our group indicated that cardioselective beta-adrenergic blockers (e.g., practolol) may be less active as inhibitors of lipolysis and, possibly, of LPL (142). We noted that propranolol administration may reduce the activity of concomitantly administered "catabolic drugs" (141).

The effects of diuretics are similar. Earlier investigations suggested that plasma total cholesterol levels might be increased following long term diuretic treatment (e.g. high doses of chlorothiazide or chlorthalidone) (138). More recently, it was shown that plasma total cholesterol may rise moderately, but that a significant increase of VLDL TG and a reduction of HDL-C may occur following treatment (143). The mechanism is totally unclear.

In contrast to the above mentioned agents, a comparative study with prazosin recently failed to indicate any specific effect on plasma lipids (144). These last studies are currently being repeated in our laboratory.

Toxic chemicals - The effect of environmental chemicals on lipid metabolism and atherosclerosis development is of topical interest. Particular attention is being devoted to agents, e.g. pesticides, which entail chronic exposure of some working categories, as well as to cigarette smoke, in view of its well established positive correlation with atherosclerosis development (145).

Chronic exposure to <u>pesticides</u> results in very typical changes of the apolipoprotein pattern. Chronically exposed workers exhibit increased HDL-C levels (146), apparently maintained throughout the exposure. Re-analysis of subjects who interrupted exposure for several years, shows a return of HDL-C to normal levels (147). The increased HDL-C following pesticides may be similar to that observed following drug treatments with potent enzyme inducers, e.g. phenobarbitone or phenytoin (148). In these cases, both increased total and HDL-C levels are noted in treated patients.

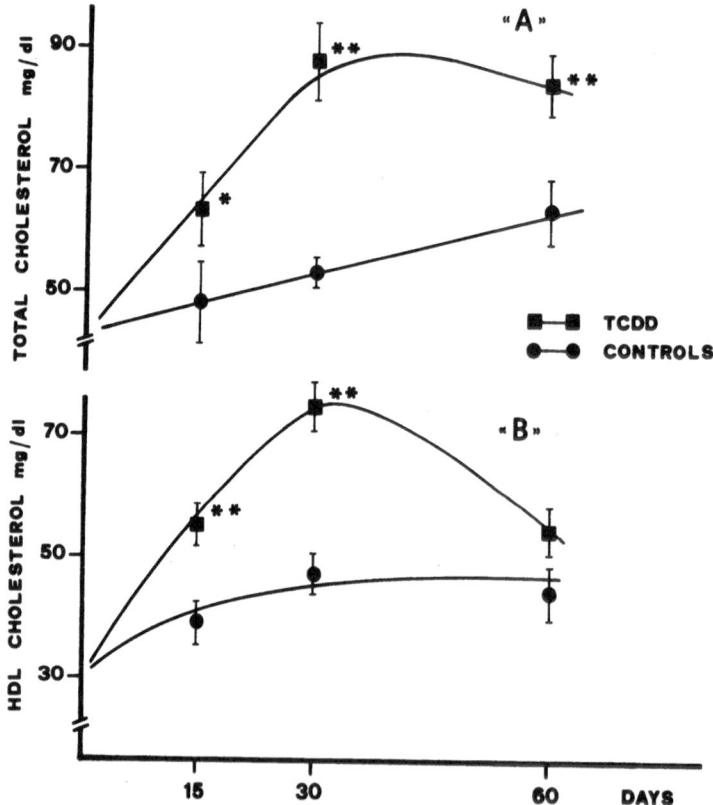

Fig. 6 – Changes of total and HDL cholesterol levels in Sprague Dawley male rats treated with a single, non lethal dose (20 γ/kg, i.p.) of tetrachlorodibenzodioxin. A long lasting, reversible increase of both biochemical parameters is noted (149).

The hypothesis that enzyme induction may be responsible for the HDL-C increases following pesticides is supported by our studies on tetrachlorodibenzodioxine (TCDD). This highly toxic chlorinated hydrocarbon, when given in single doses to rats, markedly raises total and HDL-C levels (Fig. 6) (149). The HDL increase is accompanied by apoprotein modifications consistent with the hypothesis of an increased liver secretion of HDL. The significance of these HDL changes in atherosclerosis prevention is unclear. In other studies by our group on cholesterol fed rabbits, concomitant administration of cholesterol and of TCDD appeared to increase the severity of lesions. This observation would support the hypothesis, raised by Benditt (150), that enzyme induction may favour transformation of cholesterol into toxic metabolites. On the other hand, long term studies on TCDD exposed subjects, detected a 32% reduction of vascular deaths versus the expected incidence (151).

As far as cigarette smokers are concerned, epidemiological data show decreased HDL-C levels (152). Consumption of both cigarettes and coffee may further decrease HDL-C, as above indicated. It has also been suggested that cigarette smoke may induce lipoprotein changes similar to those seen in type III individuals, by inhibiting liver clearance of chylomicron remnants, thus raising IDL levels (153). The significance of these findings in humans is yet to be established.

REFERENCES

1. C.R. Sirtori, G.C. Ghiselli, and M.R. Lovati, Dietary effects on lipoprotein composition, in: "The Lipoprotein Molecule", H. Peeters, ed., NATO Adv. Study Inst., Plenum Press, New York, p. 261 (1978).

2. G.J. Miller and N.E. Miller, Plasma high density lipoprotein concentration and development of atherosclerosis, Lancet i:16 (1978).

3. F.W. Ashley jr. and W.B. Kannel, Relation of weight changes to changes in atherogenic traits: The Framingham Study, J. Chron. Dis. 27:103 (1974).

4. J. Stamler, R. Stamler, W.F. Riedlinger, et al., Hypertension screening of 1 million Americans, J. Am. Med. Ass. 235: 2299 (1976).

5. P.C. Elwood and P.M. Sweetman, Aspirin and secondary mortality after myocardial infarction, Lancet ii:1313 (1979).

6. E. Florence and J. Quarterman, The effects of age, feeding pattern and sucrose on glucose tolerance and plasma free fatty acids and insulin concentrations in the rat, Br. J. Nutr. 28:63 (1972).

7. W.M. Bortz, P. Howat, and W.L. Holmes, The effect of feeding frequency on diurnal plasma free fatty acids and glucose levels, Metabolism 18:120 (1969).

8. R.M. Kay, S. Rao, C. Arnott, et al., Acute effects of fat ingestion on plasma high density lipoprotein components in man, Atherosclerosis 36:567 (1980).

9. D.E. Wilson and R.S. Lees, Metabolic relationships among the plasma lipoproteins: reciprocal changes in the concentrations of very low and low density lipoproteins in man. J. Clin. Invest. 51:1051 (1972).

10. B. Vessby and M. Lithell, Dietary effects on lipoprotein levels in hyperlipoproteinemia. Delineation of two subgroups of endogenous hypertriglyceridemia. Artery 1:63 (1976).

11. D.S. Goodman, F.R. Smith, A.H. Seplowitz, et al., Prediction of parameters of whole body cholesterol metabolism in humans. J. Lipid Res. 21:699 (1980).

12. C.G. Schwartz, L.G. Halloran, Z.R. Vlahcevic, et al., Preferrential utilization of free cholesterol from high density lipoproteins for biliary cholesterol secretion in man, Science 200:62 (1978).

13. P.J. Nestel and N.E. Miller, Mobilization of adipose tissue cholesterol in high density lipoprotein during weight reduction in man, in "High density Lipoproteins and Atherosclerosis", A.M. Gotto jr. et al., eds., Elsevier/ North Holland, Amsterdam, p. 51 (1978).

14. P.D. Thompson, R.W. Jeffery, R.R. Wing, and P.D. Wood, Unexpected decrease in plasma high density lipoprotein cholesterol with weight loss, Am. J. Clin. Nutr. 32:2016 (1979).

15. S.B. Hulley, R. Cohen, and G. Widdowson, Plasma high-density lipoprotein cholesterol level: influence of risk factor intervention, J. Am. Med. Ass. 238:2269 (1977).

16. F. Contaldo, P. Strazzullo, A. Postiglione, et al., Plasma high density lipoprotein in severe obesity after weight loss, Atherosclerosis 37:163 (1980).

17. D.A. Streja, E. Boyko, and S.W. Rabkin, Changes in plasma high-density lipoprotein cholesterol concentration after

weight reduction in grossly obese subjects. Brit. Med. J. 281:770 (1980).

18. J.L. Witztum, M.A. Dillingham, W. Giese, et al., Normalization of triglycerides in type IV hyperlipoproteinemia fails to correct low levels of high-density-lipoprotein cholesterol, N. Eng. J. Med. 303:907 (1980).

19. O. Turpeinen, M.J. Karvonen, M. Pekkarinen, et al., Dietary prevention of coronary heart disease: The Finnish Mental Hospital Study, Int. J. Epidemiol. 8:99 (1979).

20. J.J. Salonen, P. Puska and H. Mustanieni, Changes in morbidity and mortality during comprehensive community programme to control cardiovascular diseases during 1972-73 in North Karelia, Brit. Med. J. 2:1178 (1979).

21. J.V. Joossens, K. Vuylsteek, E. Brems-Heyns, et al., The pattern of food and mortality in Belgium, Lancet i:1069 (1977).

22. W.E. Connor, D.T. Witiak, D.B. Stone, and M.L. Armstrong, Cholesterol balance and fecal neutral steroid and bile acid excretion in normal men fed dietary fats of different fatty acid composition, J. Clin. Invest. 48:1363 (1969).

23. S.M. Grundy and E.H. Ahrens jr., The effects of unsaturated dietary fats on absorption, excretion, synthesis, and distribution of cholesterol in man, J. Clin. Invest. 49: 1135 (1970).

24. R.A.L. Sturdevant, M.L. Pearce, and S. Dayton, Increased prevalence of cholelithiasis in men ingesting a serum-cholesterol-lowering diet, N. Eng. J. Med. 288:24 (1973).

25. A. Chait, A. Onitiri, A. Nicoll, et al., Reduction of serum triglyceride levels by polyunsaturated fat. Studies on the mode of action and on the very low density lipoprotein composition. Atherosclerosis 20:347 (1974).

26. J. Shepherd, C.J. Packard, J.P. Patsch, et al., Effects of polyunsaturated fat on the properties of high density lipoproteins and the metabolism of apoprotein A-I, J. Clin. Invest. 61:528 (1978).

27. B. Vessby, J. Boberg, I.-B. Gustafsson, et al., Reduction of high density lipoprotein cholesterol and apoprotein A-I concentrations by a lipid-lowering diet, Atherosclerosis 35:21 (1980).

28. R.M. Glickman and P.H.R. Green, The intestine as a source of apoprotein A-I, Proc. Natl. Acad. Sci. U.S.A. 74:2569 (1977).

29. N.A. Spritz and M.A. Mishkel, Effects of dietary fats on plasma lipids and lipoproteins: an hypothesis for the lipid-lowering effect of unsaturated fatty acids, J. Clin. Invest. 48:78 (1969).

30. I. Hjermann, S.C. Enger, A. Helgeland, et al., The effect of dietary changes on high density lipoprotein cholesterol, Am. J. Med. 66:105 (1979).

31. R.L. Jackson, C.J. Glueck, S.N. Mathur, and A.A. Spector, Effects of diet and high density lipoprotein subfractions on the removal of cellular cholesterol, Lipids 15:230 (1980).

32. W.F. Connor, R.E. Hodges, and R.E. Bleiler, The serum lipids in men receiving high cholesterol and cholesterol-free diets. J. Clin. Invest. 40:894 (1961).

33. F.H. Mattson, B.A. Erickson, and A.M. Kligman, Effect of dietary cholesterol in man, Am. J. Clin. Nutr. 25:589 (1972).

34. H.P. Rhomberg and H. Braunsteiner, Excessive egg consumption, xanthomatosis and hypercholesterolaemia, Brit. Med. J. 2:188 (1976).

35. S. Slater, J. Mead, G.A. Dhopeshwarkar, et al., Plasma cholesterol and triglycerides in men with added eggs in the diet, Nutr. Rep Int. 14:249 (1976).

36. M.W. Porter, W. Yamamaka, S.D. Carlson, and M.A. Flynn, Effect of dietary egg on serum cholesterol and triglyceride of human males, Am. J. Clin. Nutr. 30:490 (1977).

37. M.A. Flynn, G.B. Nolph, T.C. Flynn, et al., Effect of dietary egg on human serum cholesterol and triglycerides, Am. J. Clin. Nutr. 32:1051 (1979).

38. D.C. Bronsgeest-Schoute, J.G.A.J. Hautvast, R.J.J. Hermus, et al., Dependence of the effects of dietary cholesterol and experimental conditions on serum lipids in man. I, II, III. Am. J. Clin. Nutr. 32:2183 (1980).

39. P. Mistry, A. Nicoll, C. Niehaus, et al., Effects of dietary cholesterol on serum lipoproteins in man, Prot. Biol. Fluids, 25:349 (1978).

40. R.W. Mahley, T.L. Innerarity, T.B. Bersot, et al., Alterations in human high density lipoproteins, with or without increased plasma-cholesterol, induced by diets high in cholesterol, Lancet ii:807 (1978).

41. H.D. Trowell, A.T. Southgate, T.M.S. Wolever, et al., Dietary fibre redefined, Lancet i:967 (1976).

42. D.P. Burkitt, A.R.P. Walker, and N.S. Painter, Dietary fiber and disease, J. Am. Med. Ass. 229:1068 (1974).

43. R.M. Kay and P.A. Judd, and A.S. Truswell, The effect of pectin on serum cholesterol, Am. J. Clin. Nutr. 31:562 (1978).

44. D.J.A. Jenkins, A.C. Newton, A.R. Leeds, et al., Effect of pectin, guar gum and wheat fiber on serum cholesterol, Lancet i:116 (1975).

45. T.L. Raymond, W.E. Connor, D.S. Lin, et al., The interaction of dietary fibers and cholesterol upon the plasma lipids and lipoproteins, sterol balance, and bowel function in human subjects, J. Clin. Invest. 60:1429 (1977).

46. M.R. Malinow, P. McLaughlin, L. Papworth, et al., Effect of alfalfa saponins on intestinal cholesterol absorption in rats, Am. J. Clin. Nutr. 30:2061 (1972).

47. M.R. Malinow, P. McLaughlin, H.K. Naito, et al., Effect of alfalfa meal on shrinkage (regression) of atherosclerotic plaques during cholesterol feeding in monkeys, Atherosclerosis 30:27 (1978).

48. D.J.A. Jenkins, D. Reynolds, B. Slavin, et al., Dietary fiber and blood lipids: treatment of hypercholesterolaemia with guar gum, Am. J. Clin. Nutr. 33:575 (1980).

49. A.R. Mendeloff, Dietary fibre, Presented at the First Int. Symp. on Clinical Nutrition, London, July 1980.

50. M. Stasse-Wolthuis, H.F.F. Albers, J.G.C. van Jeveren, et al., Influence of dietary fiber from vegetables and fruits, bran or citrus pectin on serum lipids, fecal lipids, and colonic function, Am. J. Clin. Nutr. 33:1745 (1980).

51. A. Ignatowski, Uber die Wirkung der tierischen Eiweisses auf die Aorta und die parenchymatosen Organer der Kaninchen, Virchows Arch. 198:248 (1909).

52. L.H. Newburgh and S. Clarkson, The production of atherosclerosis in rabbits by feeding diets rich in meat, Arch. Int. Med. 31:653 (1923).

53. S. Clarkson and L.H. Newburgh, The relation between atherosclerosis and ingested cholesterol in rabbits, J. Exp. Med. 43:592 (1926).

54. M.W. Huff, R.M.G. Hamilton, and K.K. Carroll, Plasma cholesterol levels in rabbits fed low fat, cholesterol-free, semipurified diets: effects of dietary proteins, protein hydrolysates and amino acid mixtures, Atherosclerosis 28:187 (1977).

55. C.R. Sirtori, E. Agradi, F. Conti, et al., Soybean protein diet in the treatment of type II hyperlipoproteinaemia, Lancet i:275 (1977).

56. C.R. Sirtori, G.C. Descovich, and G. Noseda, Textured soy protein and serum cholesterol, Lancet i:149 (1980).

57. J. Van Raaj, M.B. Katan, and J.G.A.J. Hautvast, Casein, soja protein and serum cholesterol, Lancet ii:958 (1979).

58. C.R. Sirtori, E. Gatti, O. Mantero, et al., Clinical experience with the soybean protein diet in the treatment of hypercholesterolemia, Am. J. Clin. Nutr. 32:1645 (1979).

59. G.C. Descovich, C. Ceredi, A. Gaddi, et al., Multicenter study of soybean protein diet for outpatient hypercholesterolaemic patients, Lancet ii:709 (1980).

60. W.L. Holmes, G.B. Rubel, and S.S. Hood, Comparison of the effects of dietary meat versus dietary soybean protein on plasma lipids of hyperlipidemic individuals, Atherosclerosis 36:379 (1980).

61. C.R. Sirtori, G.C. Descovich and G. Noseda, The soybean protein diet does not lower plasma cholesterol?, Atherosclerosis, in press (1981).

62. R. Fumagalli, R. Paoletti and A. Howard, Hypocholesterolaemic effect of soya, Life Sci. 22:947 (1978).

63. D.M. Kim, K.T. Lee, J.M. Reiner, and W.A. Thomas, Effects of a soy protein product on serum and tissue cholesterol concentrations in swine fed high-fat, high cholesterol diets, Exp. Mol. Pathol. 29:385 (1978).

64. J.M. Potter and P.J. Nestel, Greater bile acid excretion with soybean than with cow milk in infants, Am. J. Clin. Nutr. 32:1645 (1976).

65. R. Fumagalli, L. Soleri, O. Mantero, et al., Cholesterol balance studies in type II hypercholesterolaemic patients treated with the soybean protein diet, Clin. Sci., submitted for publ.

66. J.H. Mendelson and N.K. Mello, Alcohol-induced hyperlipidemia and beta-lipoproteins, Science 180:1372 (1973).

67. L. Puglisi, V. Caruso, F. Conti, et al., Effect of hypolipidemic and hypoglycemic drugs on ethanol induced hypertriglyceridemia in rats, Pharmacol. Res. Comm. 9:71 (1977).

68. D.E. Wilson, P.H. Schreibman, A.C. Breston, and R.A. Arky, The enhancement of alimentary lipemia by ethanol in man, J. Lab. Clin. Med. 75:264 (1970).

69. W.P. Castelli, T. Gordon, M.C. Hjortland, et al., Alcohol and

blood lipids. The cooperative lipoprotein phenotyping study, Lancet ii:153 (1977).

70. G.J. Miller and N.E. Miller, Plasma high-density lipoprotein concentration and development of ischaemic heart-disease, Lancet i:16 (1975).

71. K. Yano, G.G. Rhoads, and A. Kagan, Coffee, alcohol and risk of coronary death, among Japanese men living in Hawaii, N. Eng. J. Med. 297, 405 (1977).

72. A.S. St. Leger, A.L. Cochrane, and F. Moore, Factors associated with cardiac mortality in developed countries with particular reference to the consumption of wine, Lancet i:1017 (1979).

73. P. Ducimetiere, J.L. Richard, F. Cambien, et al., Coronary heart disease in middle-age Frenchmen, Lancet, i:1347 (1980).

74. P. Nilsson-Ehle, Alcohol induced alterations on lipoprotein lipase activity and plasma lipoproteins, in "Metabolic Effects of Alcohol", P. Avogaro et al., eds., Elsevier/ North Holland, Amsterdam, p. 175 (1979).

75. J.R. Patsch, A.M. Gotto Jr., T. Olivecrona, and S. Eisenberg, Formation of high density lipoprotein$_2$-like particles during lipolysis of very low density lipoproteins in vitro, Proc. Natl. Acad. Sci. USA 75:4519 (1978).

76. T.R. Dawber, W.B. Kannel, and T. Gordon, Coffee and cardiovascular disease: Observations from the Framingham study, N. Eng. J. Med. 291:871 (1974).

77. D. Robertson, J.L. Frohlich, R.K. Carr, et al., Effects of caffeine on plasma renin activity, catecholamines and blood pressure, N. Eng. J. Med., 298:181 (1978).

78. S. Hayden, H.A. Tyroler, G. Heiss, et al., Coffee consumption and mortality, J. Am. Med. Ass., 138:1472 (1978).

79. N.E. Hag, J.G. Elliott, and P.A. Lachance, Effect of tea on serum cholesterol using a hypercholesterolemic rat model, Nutr. Rep. Int., 15:89 (1977).

80. M.N. Callahan, M.N. Rohovsky, R.S. Robertson, and D.W. Yesair, The effect of coffee consumption on plasma lipids, lipoproteins, and the development of aortic atherosclerosis in Rhesus monkeys fed an atherogenic diet, Am. J. Clin. Nutr. 32:834 (1979).

81. S. Heyden, G. Heiss, C. Manegold, et al., The combined effect of smoking and coffee drinking on LDL and HDL cholesterol. Circulation, 60:22 (1979).

82. A cooperative trial in the prevention of ischaemic heart disease using clofibrate, Brit. Heart J., 40:1069 (1978).

83. W.H.O. Cooperative trial on primary prevention of ischaemic heart disease using clofibrate to lower serum cholesterol: mortality follow-up, Lancet ii:379 (1980).

84. E.A. Nikkila, J.K. Huttunen, and C. Ehnholm, Effect of clofibrate on post-heparin plasma triglyceride lipase activities in patients with hypertrigliceridemia, Metabolism 26:179 (1977).

85. G. Klose, J. Behrendt, J. Vollmar, and H. Greten, Wirkung von Bezafibrate auf die Lipoprotein Lipase Aktivitat und die Leber-Triglyceridhydrolase bei gesunden Versuschpersonen, in "Lipoproteine und Herzinfarkt", H. Greten et al., G. Witzstrock Verlag, Baden-Baden, p. 185 (1979).

86. L.A. Carlson, A.G. Olsson, and D. Ballantyne, On the rise of low density and high density lipoproteins in response to the treatment of hypertriglyceridemia in type IV and V hyperlipoproteinemias, Atherosclerosis 26:603 (1973).

87. G. D'Atri, P. Gomarasca, E. Galimberti, et al., Clofibrate, Pirinixil (BR 931) and Wy-14,643 do not affect body cholesterol in Sprague-Dawley rats, Atherosclerosis 37: 475 (1980).

88. S.M. Grundy, E.H. Ahrens, G. Salen, et al., Mechanism of action of clofibrate on cholesterol metabolism in patients with hyperlipidemia, J. Lipid Res. 13:531 (1972).

89. G. Schlierf, M. Chwat, E. Feuerborn, et al., Biliary and plasma lipids and lipid-lowering chemotherapy. Atherosclerosis 36:323 (1980).

90. T.A. Miettinen, J.K. Huttunen, C. Ehnholm, et al., Effect of long-term antihypertensive and hypolipidemic treatment on high density lipoprotein and apolipoproteins A-I and A-II, Atherosclerosis 36:249 (1980).

91. J.G. Brook, A. Lavy, M. Aviram, and M. Zinder, The concentration of high density lipoprotein in patients with type IV hyperlipoproteinemia and the effect of clofibrate, Atherosclerosis 36:461 (1980).

92. The Coronary Drug Project: Clofibrate and niacin in coronary heart disease, J. Am. Med. Ass. 231:360 (1975).

93. J. Reddy, D.L. Azarnoff, and C.R. Sirtori, Hepatic peroxisome proliferation: induction by BR-931, a hypolipidemic analog of WY-14,643, Arch. Int. Pharmacodyn. 234:4 (1978).

94. J. Reddy, D. Azarnoff and C. Hignite, Hypolipidemic hepatic peroxisome proliferators form a novel class of hepatocar-

cinogens, Nature 283:397 (1980).

95. M. Hanefeld, Ch. Remmer, W. Leonhardt, et al., Effects of p-chlorophenoxy-isobutyric acid (CPIB) on the human liver, Atherosclerosis 36:159 (1980).

96. H. Ishii, N. Fukumori, S. Horie, and T. Suga, Effects of fat content in the diet on hepatic peroxisomes of the rat, Biochim. Biophys. Acta 617:1 (1980).

97. H. Ishii, and T. Suga, Clofibrate-like effects of acetylsalicylic acid on peroxisome and on hepatic and serum triglyceride levels, Biochem. Pharmacol. 28:2829 (1979).

98. D.L. Azarnoff, J. Reddy, C. Hignite, and T. Fitzgerald, Structure activity relationships of clofibrate like compounds on lipid metabolism, in "Proc. VI Int. Congress of Pharmacology", vol. 4, M. Vapaatalo, ed., Pergamon Press, Oxford, p. 137 (1976).

99. E.A. Nikkila and O. Pykalisto, Induction of adipose tissue lipoprotein lipase by nicotinic acid, Biochim. Biophys. Acta 152:421 (1968).

100. L.A. Carlson and L. Oro, Acute effect of a sustained release pyridyl carbinol preparation, ronicol retard, on plasma free fatty acids, Acta Med. Scand. 183:457 (1968).

101. T. Langer and R.T. Levy, The effect of nicotinic acid on the turnover of low density lipoproteins in type II hyperlipoproteinemia, in "Metabolic effects of nicotinic acid and its derivatives", K.F. Gey and L.A. Carlson, eds., Hans Huber Publ., Bern, p. 641 (1971).

102. J. Shepherd, C.J. Packard, J.R. Patsch, et al., Effects of nicotinic acid therapy on plasma high density lipoprotein subfraction distribution and composition, and of apolipoprotein A metabolism. J. Clin. Invest. 63:858 (1979).

103. C.J. Packard, J.M. Stewart, J.L. Third, et al., Effects of nicotinic acid therapy on high-density lipoprotein metabolism in type II and type IV hyperlipoproteinemia, Biochim. Biophys. Acta 618:53 (1980).

104. H.Y.I. Mok and S.M. Grundy, Effects of nicotinic acid on biliary lipids in man, Gastroenterology 73:1235 (Abs.) (1977).

105. B. Angelin, K. Einarsson, and B. Lleijd, Biliary lipid composition during treatment with different hypolipidaemic drugs, Eur. J. Clin. Invest. 9:185 (1979).

106. G. Baggio, G. Briani, R. Fellin et al., Effect of tiadenol on the concentration and composition of serum lipoproteins in familial hypercholesterolemia, Artery 5:486

(1979).

107. J. Rouffy and J. Loeper, Effects hypolipidemiants du bis
     (hydroxy-2-ethylthio) 1.10 décane (LL 1558), Thérapie 27:
     433 (1974).

108. E. Martin and G. Feldman, Etude hystologique et ultrastruc-
     tural du foie chez le rat aprés administration subaigue
     d'un nouvel agent hypolipémiant, le bis (hydroxyethyl-
     thio) 1.10 décane, Pathol. Biol. XXII-II:179 (1974).

109. D. Kritchevsky, S.A. Tepper, S.K. Czarnecki, et al., Influ-
     ence of tiadenol, bis-(hydroxy-ethyltio) 1.10 decane, on
     cholesterol metabolism in rats, Pharmacol. Res. Comm.
     11:475 (1979).

110. G. Franceschini, A. Catapano, A. Poli, et al., Pharmacologi-
     cal studies on tiadenol in type IV patients: evidence for
     a mechanism of action different from other lipid lowering
     drugs, Atherosclerosis, submitted for publ.

111. G. Schonfeld, F. Bell and D.H. Alpers, Intestinal apopro-
     teins during fat absorption, J. Clin. Invest. 61:1539
     (1977).

112. C.R. Sirtori, A. Catapano, G.C. Ghiselli, et al., Metformin:
     an antiatherosclerotic agent modifying very low density
     lipoproteins in rabbits, Atherosclerosis 26:79 (1977).

113. C.R. Sirtori, A. Catapano, G.C. Ghiselli, et al., Effects of
     metformin on lipoprotein composition in rabbits and man,
     Prot. Biol. Fluids 25:379 (1978).

114. R.W. Stout, J.D. Brunzell, D. Porte Jr., and E.L. Bierman,
     Effect of phenformin on lipid transport in hypertriglycer-
     idemia, Metabolism 23:815 (1974).

115. W. Lintz, W. Berger, W. Aenishaenslin et al., Butylbiguanide
     concentration in plasma, liver and intestine after intra-
     venous and oral administration to man. Eur. J. Clin. Phar-
     macol. 7:433 (1974).

116. G. Weber, A. Catapano, G.C. Ghiselli and C.R. Sirtori, Exper-
     imental studies on the antiatherosclerotic effect of
     metformin, in "International Conference on Atherosclero-
     sis", L.A. Carlson et al., Eds., Raven Press, New York,
     p. 319 (1978).

117. R. Paoletti, R. Fumagalli, A. Poli, and C.R. Sirtori, Bioche-
     mical analysis of hypolipidemic versus anti-atherosclero-
     tic drugs, Atherosclerosis Rev. 7:213 (1980).

118. C.R. Sirtori, G. Franceschini, M. Sirtori, et al., Apopro-
     tein changes following treatments with hypolipidemic
     drugs, in "Atherosclerosis V", A.M. Gotto et al., eds.,

Springer Verlag, New York, p. 90 (1980).

119. G.C. Descovich, U. Montaguti, C. Ceredi, et al., Long-term treatment with metformin in a large cohort of hyperlipidemic patients, Artery 4:348 (1978).

120. E. Tremoli, G.C. Ghiselli, P. Maderna, et al., Metformin reduces platelet hypersensitivity in hypercholesterolemic rabbits, Atherosclerosis, submitted for publ.

121. C.J. Glueck, S. Ford, D. Scheel, and P. Steiner, Colestipol and cholestyramine resin, J. Am. Med. Ass. 222:676-681 (1972).

122. A.E. Dorr, K. Gundersen, J.C. Schneider Jr., et al., Colestipol hydrochloride in hypercholesterolemic patients. Effects on serum cholesterol and mortality, J. Chron. Dis. 31:5 (1978).

123. R.W. Wissler, D. Vesselinovitch, J. Borensztajn, and R. Hughes, Regression of severe atherosclerosis in cholestyramine-treated Rhesus monkeys with or without a low-fat, low-cholesterol diet, Circulation 52 (suppl. II): II-16 (1976).

124. R.G. De Palma, E.M. Bellon, S. Koletsky, and D.L. Schneider, Atherosclerotic plaque regression in Rhesus monkeys induced by bile acid sequestrant, Exp. Mol. Pathol. 31:423 (1979).

125. C.D. Moutafis, N.B. Myant, M. Mancini, and P. Oriente, Cholestyramine and nicotinic acid in the treatment of familial hyperbetalipoproteinemia in the homozygous form, Atherosclerosis 20, 105 (1971).

126. G.S. Boyd, A.M. Grimwade, and M.E. Lawson, Studies on microsomal cholesterol 7α-hydroxylase, Eur. J. Biochem. 37:334 (1973).

127. J.L. Witztum, G. Schonfeld, and S.W. Weidman, The effects of colestipol on the metabolism of very-low-density lipoproteins in man, J. Lab. Clin. Med. 88:1008 (1976).

128. R. Adler, E. Margules, R. Motson, et al., Increased production of triglyceride-rich lipoproteins after partial biliary diversion in the Rhesus monkey, Metabolism 27:607 (1978).

129. B. Angelin, K. Einarsson, and B. Leijd, Effect of chenodeoxycholic acid on serum and biliary lipids in patients with hyperlipoproteinemias, Clin. Sci. 54:451 (1978).

130. G. Sigurdsson, A. Nicoll, and B. Lewis, Conversion of very low density lipoprotein to low density lipoprotein. A metabolic study of apolipoprotein B kinetics in normal

subjects, J. Clin. Invest. 56:1481 (1975).

131. A. Soutar, N.B. Myant, and G.R. Thompson, Simultaneous measurement of apolipoprotein B turnover in very-low and low-density lipoproteins in familial hypercholesterolemia, Atherosclerosis 28:247 (1977).

132. G.C. Ghiselli, M. Marinovich, and C.R. Sirtori, Apoprotein B turnover in normal rabbits: evidence for a dual biosynthetic and catabolic pool of apo B in low density lipoproteins, Artery, submitted for publication.

133. R.I. Levy and T. Langer, Hypolipidemic drugs and lipoprotein metabolism, Adv. Exp. Med. Biol. 26:155 (1972).

134. J. Shepherd, C.J. Packard, S. Bicker, et al., Cholestyramine promotes receptor-mediated low-density lipoprotein metabolism, N. Eng. J. Med. 302:1219 (1980).

135. H.R. Slater, C.J. Packard, S. Bicker, and J. Shepherd, Effects of cholestyramine on receptor-mediated plasma clearance of human low density lipoproteins in the rabbit, J. Biol. Chem. 255:10210 (1980).

136. J.L. Witztum, G. Schonfeld, S.W. Weidman, et al., Bile sequestrant therapy alters the compositions of low-density and high-density lipoproteins, Metabolism 28:221 (1979).

137. N. Tanaka, S. Sakaguchi, K. Oshige et al., Effect of chronic administration of propranolol on lipoprotein composition, Metabolism 25:1071 (1976).

138. R.P. Ames and P. Hill, Elevation of serum lipid levels during diuretic therapy of hypertension, Am. J. Med. 61:748 (1976).

139. J. Shaw, J.D.F. England, K. Oshige, et al., Beta-blockers and plasma triglycerides, Brit. Med. J. i:986 (1978).

140. P. Bielmann and G. Leduc, Effect of metaprolol and propranolol on lipid metabolism, Int. J. Clin. Pharmacol. Biopharm. 17:378 (1979).

141. G. Noseda and C.R. Sirtori, Klinische Prufung von "Tibric Acid", einem neuem lipidsenkenden Stoff, Schweiz. Med. Wschr. 104:1917 (1974).

142. C.R. Sirtori, D.W. Shoeman, and D.L. Azarnoff, Dissociation of the metabolic and cardiovascular effects of the β-adrenergic blocker practolol, Pharmacol. Res. Comm. 4:123 (1972).

143. C. Joos, H. Kewitz, and D. Reinholdt-Kourniati, Effects of diuretics on plasma lipoproteins in man, Eur. J. Clin.

Pharmacol. 17:251 (1980).

144. P. Leren, P.O. Foss, A. Helgeland, et al., Effect of propranolol and prazosin on plasma lipids, Lancet ii:4 (1980).

145. K. Ball and R. Turner, Smoking and the heart. The basis for action, Lancet ii:822 (1974).

146. L.A. Carlson and B. Kolmodin-Hedman, Hyper-α-lipoproteinemia in men exposed to chlorinated hydrocarbon pesticides, Acta Med. Scand. 192:29 (1972).

147. L.A. Carlson and B. Kolmodin-Hedman, Decrease in α-lipoprotein cholesterol in men after cessation of exposure to chlorinated hydrocarbon pesticides, Acta Med. Scand. 201: 375 (1977).

148. E.A. Nikkila, M. Koste, C. Ehnholm, and J. Viikari, Elevation of high density lipoprotein in epileptic patients treated with phenytoin, Acta Med. Scand. 204:517 (1978).

149. A. Poli, G. Franceschini, L. Puglisi, and C.R. Sirtori, Increased total and high density lipoprotein cholesterol, with apolipoprotein changes resembling streptozotocin diabetes in tetrachlorodibenzodioxin (TCDD) treated rats, Biochem. Pharmacol. 29:835 (1980).

150. M.R. Juchau, J.A. Bond, and E.A. Benditt, Aryl 4-monoxygenase and cytochrome P-450 in the aorta: Possible role in atherogenesis, Proc. Natl. Acad. Sci. USA 73:3723 (1976).

151. J.A. Zack and R.R. Suskind, The mortality experience of workers exposed to tetrachlorodibenzodioxin in a trichlorophenol process accident, J. Occup. Med. 22:11 (1980).

152. R.J. Garrison, W. Kannel, M. Feinleib, et al., Cigarette smoking and HDL cholesterol. The Framingham Offspring Study, Atherosclerosis 30:17 (1978).

153. R.S. Gardner, D.L. Topping, and P.A. Mayes, Immediate effects of carbon monoxide on the metabolism of chylomicron remnants by perfused rat liver, Biochim. Biophys. Res. Comm. 82:526 (1978).

ENDOGENOUS AGENTS IN PLATELET THROMBOSIS

G.V.R. Born and M.A.A. Kratzer

Department of Pharmacology, University of London

King's College, London WC2R 2LS, United Kingdom

CLINICAL AND PATHOLOGICAL FEATURES OF ARTERIAL THROMBOSIS

Any hypothesis for arterial thrombosis must be able to account for the following facts: (1) thrombi do not form in normal arteries; (2) thrombi form in atherosclerotic arteries; (3) arterial thrombi consist initially of aggregated platelets; (4) atherosclerosis increases slowly, whereas thrombosis occurs rapidly and is individually unpredictable. Therefore, atherosclerotic arteries must be subject to sudden, unpredictable events capable of initiating platelets aggregation; (5) most occlusive thrombi are associated with fissures in underlying atheromatous placques.

The erythrocyte-haemodynamics hypothesis (Born 1979) proposes that the sudden unpredictable event that starts arterial (typically coronary) thrombosis is placque fissure; haemorrhage through the fissure is associated with increased haemodynamic stress causing ADP (and other adenine nucleotides) to appear in the plasma; and this ADP is principally responsible for activating platelets and their aggregation as mural thrombi.

Much progress has been made with support for this hypothesis (Born et al.,1976;Born & Wehmeier,1979).

Evidence includes the following: (1) in atherosclerotic arteries platelet thrombi form only when blood flow is sufficiently abnormal, i.e., as a result of haemorrhage into a fissure, or in tortuous and or stenotic regions; (2) in artificial blood vessels mural thrombi of platelets grow where, and only where, flow

is non-laminar; (3) in artificial vessels the formation of platelet
thrombi in non-laminar flow depends on the presence of red cells;
(4) adhesion and aggregation of platelets on artificial surfaces
increase with red cell concentration and are abolished by ADP-re-
moving enzymes; (5) the haemodynamic stress associated with experimental haemorrhage is insufficient in duration and magnitude to
activate platelets directly, but sufficient in both to induce
release of red cell ADP (Born & Wehmeier 1979).

    The thrombogenic adhesion of platelets to vessel walls, there-
fore, depends indirectly on the haemodynamic properties of the
blood as it flows through arteries constricted and/or fissured by
atherosclerosis.

    Gross and histological appearances of arterial thrombi estab-
lish that their central mass consists mainly of aggregated platelets.
What, therefore, is the mechanism responsible for rapid and extensive platelet aggregation in an artery as an apparently random event
in time ? Close serial sectioning of obstructed coronary arteries
established some time ago that the platelet thrombus responsible is
usually if not invariably associated with recent haemorrhage into
an underlying atherosclerotic placque (Friedman & Byers 1965; constantinides 1966; Davies 1981). The haemorrhages occur through
fissures or fractures in the placque; and it is a reasonable as-
sumption that the sudden appearance of such a fissure or fracture
is the random, individually unpredictable event affecting coronary
arteries that has to be assumed to account for the clinical onset
of acute myocardial infarction (Born 1979). Why such a defect
should develop at a particular moment is uncertain. Perhaps it is
analogous to the sudden appearance of fine cracks in the wings of
jet aircraft which can be ascribed to nothing more precise than
the cumulative effects of variable stresses on metal known as metal
fatigue. The chance event of placque fissure can in principle be
prevented only by preventing atherosclerosis which is, as we know,
still very problematical. Fortunately, the subsequent thrombotic
process due to platelet aggregation is now understood to the extent
that it may become preventable by drugs before long.

    How does haemorrhage into a ruptured placque start off plate-
let thrombogenesis ? This can be regarded as part of the general
question of how platelets are caused to aggregate through haemorrhage, most effectively from arteries. An explanation commonly
put forward is that the process is initiated by platelets adhering
to collagen which is exposed where damaged vessel walls are denuded
of endothelium (Mustard,Moore,Packham & Kinlough-Rathbone 1977;

Packham & Mustard 1977). Adhering platelets then release other agents including thromboxane A_2 and ADP which in turn are responsible for the adhesion of more platelets as growing aggregates. This explanation is unlikely to be correct for the following reasons. First, haemostatic and thrombotic aggregates of platelets grow without delay and very rapidly (Hughes 1959). When an arterial 200 μm in diameter is cut into laterally, the rate of accession of platelets to the haemostatic plug is of the order 10^4 per second in the first seconds (Born & Richardson 1981). In contrast, the adhesion of platelets to collagen begins, even under optimal conditions for rapid reactivity, only after a delay or lag period of several seconds (Wilner et al. 1968). Secondly, platelets tend to aggregate as mural thrombi when anticoagulated blood flows through (Didisheim,Pavlovsky & Kobayashi 1972) plastic vessels, for example, in artificial organs such as oxygenators or dialysers (Richardson,Galletti & Born 1976) which contain no collagen nor anything else capable of activating platelets similarly. This implies that there are conditions under which platelets are activated in the blood by something other than collagen or other constituents of the walls of living vessels. The placque on which a thrombus grows has usually narrowed the arterial lumen. At constant blood pressure the flow of blood is faster through the constriction than elsewhere in the artery. Therefore, high flow and wall shear rates are no hindrance to the aggregation of platelets as thrombi (Born 1977). Indeed, the question arises of whether the activation of platelets which precedes their aggregation depends in some way on abnormal haemodynamic conditions.

Measurements of the haemodynamic forces required to activate platelets directly (Hellums & Brown 1977) indicated that the blood flow over atherosclerotic lesions in vivo is unable to do so (Colantuoni et al. 1977). Therefore, the activation must be indirected. Now it has been known for many years that platelets can be activated by at least one agent, namely ADP, derived from the red cells which outnumber and surround the platelets in the blood (Gaarder,Jonsen,Laland,Hellem & Owren 1961).

Clear evidence of increased platelet adhesiveness brought about by the operation of low-mechanical factors on erythrocytes was provided by experiments in which blood was made to flow through branching channels in extra-corporeal shunts (Rowntree & Shionya 1927; Mustard et al. 1962). Deposits of platelets formed consistently on the shoulders of a bifurcation in the flow chamber but nowhere else in the channels. In such divergent flow situations there is

boundary layer separation which is accompanied by flow delays or
stasis (Fox & Hugh 1966). This might by itself be expected to increase the probability of platelet aggregation. However, when the
chambers were perfused not with blood but with platelet-rich plasma,
no deposit was formed showing that red cells were also essential.

The dependence of the deposition of platelets from flowing
blood on the presence of red cells could be caused by physical or
chemical mechanisms or, of course, by both acting synergistically.
A physical mechanism would depend essentially on an increase in the
diffusivity of platelets caused by the flow behaviour of the erythrocytes. Indeed, the diffusivity of platelets in flowing blood
has been estimated to be two orders of magnitude greater than that
calculated for platelets diffusing in plasma (Turitto et al. 1972;
Turitto & Baumgartner 1975). This is consistent with the enhanced
radial fluctuations of erythrocytes or of latex microspheres (2μm
in diameter) in flowing suspensions of red cell ghosts (Goldsmith
1971). High platelet diffusivity is required also to explain the
growth of mural thrombi. This must depend on successful platelet-
to-platelet collisions, the rate of which between platelets following streamlines near the walls would hardly be sufficiently high
to account for the rapidity of growth observed in vivo (Begent &
Born 1970; Richardson 1973).

IS ADP INVOLVED IN ARTERIAL THROMBOGENESIS ?

There is increasing evidence for a chemical mechanism in the
increased adhesiveness of platelets in presence of red cells, i.e.,
through their ADP. The concentrations of ADP required for activating platelets are small (10^{-6} M or less) and ADP is also rapidly
dephosphorylated in blood, so that its direct demonstration there
under conditions relevant to thrombogenesis is difficult.

It has recently become possible to demonstrate the appearance
of free ADP in blood directly in concentrations sufficiently high
to activate platelets (Schmid-Schönbein et al. 1979). In specially
designed apparatus whole blood or resuspended red cells are exposed
to controlled, different shear stresses for know time periods.
The apparatus is designed to cover the range of these variables
presumed to be relevant to the in vivo situations. The experiments
show that ADP appears in the plasma in concentrations required for
platelet activation (0.1 to 1.0 μM) but in direct proportion to
free haemoglobin, indicating that platelet activation can result
from small degrees of haemolysis due to haemodynamic stresses such
as occur during haemorrhage, whether external or through a placque
fissure.

It is not yet certain whether the appearance of free ADP is
rapid enough to account for in vivo aggregation. This process
appears faster than the release of ADP from the platelets themselves or of thromboxane A_2 produced by them which, in any case,
induces aggregation via ADP (personal communications from B. Samuelsson and A. Marcus).

When blood vessels are injured so that they bleed, circulating platelets adhere to the damaged vessel wall and aggregate within
the first seconds. The mechanism of the initial platelet aggregation remains uncertain. To investigate the initiation stage of
haemostasis the carotid arteries of rats were punctured with a
100 μm needle and free ATP, as an indicator of ADP, was measured in
the emerging blood (Kratzer & Born 1981). This was brought into
contact with luciferin-luciferase in a polyethylene tube, internal
diameter 0.8 mm. The light produced at the blood/enzyme interface
was measured with a sensitive photon-counting device which gave
background counts of 1 photoelectron/sec. and could detect 10^{-8} M ATP
in 2 μl blood.

When an artery was injured, the emerging blood contained about
10^{-7} M ATP in a first peak after about 2 sec. After about one min. the
ATP concentration rose to a second peak of about 5×10^{-6} M. This
was decreased by heparin (500 U/kg body weight) or by chlorpromazine (1 mg/kg). The observations suggest that the second peak represents ATP released from platelets. The source of ATP accounting
for the first peak remains uncertain; possibly this ATP is released from red cells undergoing high shear stress from the haemodynamic
effects of haemorrhage.

Our observations suggest a new approach to the prophylaxis
of arterial, e.g., coronary thromboses (Born 1979) . This approach
would require the demonstration that their incidence is diminished
by drugs which, in clinically acceptable blood concentrations, do
not inhibit the release of activating agent, presumably ADP, from
red cells during rheological stresses such as occur in potentially
thrombogenic arteries. Such a demonstration may then also explain
the effect of dipyridamole or sulfinpyrazone in preventing increased utilisation of circulating platelets under potentially thrombogenic conditions. This cannot easily be accounted for by any direct
action on platelets by either drug at its clinically effective concentration. Perhaps these drugs act mainly on the red cells to
diminish their activating effect on platelets.

Some time ago I had the idea that drugs capable of counteract-
ing haemolysis (Seeman 1972) might diminish this activating effect
of red cells on platelets and so inhibit their aggregation as throm-

bi. Experimental evidence for this idea came with the demonstration (Born,Bergqvist & Arfors 1976) that chlorpromazine added to
anticoagulated human blood in concentrations which in vitro diminish
hypotonic haemolysis but have no direct effect on platelet aggregation (Mills & Roberts 1967) prolong the bleeding time from small
holes in artificial vessels where extravasation is terminated, as
in vivo, by aggregated platelets. More recent experiments (Born &
Wehmeier 1979) support the conclusion that this effect is accounted
for by the antihaemolytic action of chlorpromazine. This has led
to the suggestion (Born 1976; Born 1979a; 1979b) that other drugs
possessing this effect of chlorpromazine may diminish the incidence
of arterial, particularly of coronary, thrombosis when it is induced by conditions of abnormal stress on the red cells, such as
through haemorrhage into atheromatous lesions.

As already proposed (Born 1979b), evidence for or against
this proposition could perhaps be obtained by comparing the incidence of acute coronary occlusions in populations on longterm treatment with chlorpromazine (or other drugs acting in this respect
like chlorpromazine) with the incidence in control populations not
on such drugs. The only conceivably relevant evidence of which I
have been made aware up to now (through the courtesy of dr. J.A.
Baldwin, Director of the Oxford University Unit of Clinical
Epidemiology) is an investigation of mortality in Norwegian psychiatric hospitals during the period 1950 to 1962 (Odegard 1967).
This concluded that mortality from circulatory disease, predominantly "coronary disease" and "infarction", was higher in the
mental hospital population than in general population, although
the excess was not as much as that from most other causes. Within
the patient population, the excess mortality from coronary disease
and infarction was less for schizophrenics than for all other psy-
choses. Furthermore, the excess mortality from circulatory disease diminished strikingly after 1957, particularly when compared
with the period 1926-41, because the mortality did not rise to
the same extent in the hospital as in the general population.

These conclusions, if confirmed, are of course open to
different interpretations. Epidemiological considerations apart,
chlorpromazine has many effects in the body, and also a large
number of metabolites. Furthermore, the concentration of chlorpromazine in patients' plasma (Mackay,Healey & Baker 1974) is one
to two orders of magnitude lower than that required to prolong the
ex-vivo bleeding time described below. Therefore, if patients'
blood were used for determining this bleeding time, a prolongation

might be expected only if the drug or a similarly active metabolite were concentrated in red cell membranes. On the other hand, it has been observed that single clinical doses (5-20 mg) of chlorpromazine injected intramuscularly into apparently healthy volunteers cause the Ivy bleeding time to be significantly prolonged (Zahavi & Schwartz 1978).

Our experimental observations with chlorpromazine make it attractive to suggest that the general introduction of this drug for the control of schizophrenic inpatients from about 1955 onwards accounts for their relative protection against cardiac mortality at a time when it was increasing rapidly in Norway and elsewhere, including Britain. It would be interesting to learn of other information which may support or, just as important, invalidate this line of thought. It may be worthwhile to investigate appropriate populations from this point of view.

ARE PROSTAGLANDINS INVOLVED IN ARTERIAL THROMBOGENESIS ?

A different explanation of arterial thrombosis has been widely canvassed (Gryglewsky et al. 1976). This may be referred to conveniently as the "prostacyclin/thromboxane hypothesis", and it proposes that: (1) whether or not thrombosis occurs in arteries depends on a balance betweeen prostacyclin (PGI_2) which inhibits platelets and thromboxane A_2 (TXA_2) which aggregates platelets; (2) in normal arteries thrombosis is prevented by prostacyclin in the blood where its sources are lungs and endothelium; (3) part of the "endothelial" prostacyclin originates in endoperoxides transferred from adherent platelets; (4) the walls of atherosclerotic arteries synthesise less prostacyclin, and those of artificial blood vessels none at all. Platelets endoperoxides are therefore diverted to thromboxane A_2 production, which is responsible for activating the platelets and their aggregation as thrombi.

In view of the extraordinary inhibitory potency of prostacyclin, this hypothesis appears attractive, but it is not enterely satisfactory because it does not explain: (1) how platelets can produce endoperoxides for utilisation by endothelium without the simultaneous formation of thromboxane A_2 by the platelets which should aggregate them; (2) that platelet adhesion to vessel walls can be seen to occur under conditions in which it might be expected to be prevented by endothelial prostacyclin; (3) why platelet thrombi do not form in artificial blood vessels, i.e., in the

absence of prostacyçlin, <u>except</u> under particular haemodynamic
conditions <u>and</u> in the presence of red cells; (4) why arterial
thrombosis doeɐ not occur continously on all atherosclerotic
lesions; (5) the essential clinical characteristics of arterial
thrombosis, viz, its unpredictable and sudden onset, commonly
in patients with either longstanding or minimal atherosclerotic
lesions.

       Information about the mode of origin of prostaglandins and
related agents doesᵉ suggest a mechanism by which they could be
brought into thrombogenesis through haemodynamic effects. Platelets
themselves produce the aggregating agent thromboxane A_2 (Hamberg
et al. 1975; Svensson et al. 1976). The first step in its formation is the release of arachidonic acid from phospholipids in the
cell membrane catalysed by the enzyme phospholipase A_2 which is
normally inactive; how the enzyme is activated phsyiologically
is not known. Perhaps activation is initiated by small distortions
of the outer membrane of platelets when they pass through regions
in which the haemodynamic forces are greater than in the normal
circulation, as during haemorrhage into a ruptured atherosclerotic
placque. This fluid-mechanical activation of platelets would not
involve red cells which apparently do not contain a thromboxane-
forming system.

IS COLLAGEN INVOLVED IN ARTERIAL THROMBOGENESIS ?

       There has been ample confirmation of the observation (Di-
disheim,Pavlovsky & Kobayashi 1972) that "thrombosis induced
mechanically in a teflon shunt appears indistinguishable from
thrombosis induced by mechanical or electrical injury in living
vessels, despite the absence of endothelium, collagen and muscle
from the teflon tubing".
       These observations do not support the assertion that haemo-
static plugs or thrombi are initiated through the collision of
platelets with exposed collagen. On the other hand it has recently
been shown that a nonapeptide sequence of collagen is a specific
binding site for platelets (Legrand,Karnignian,Le Francier,Vauvel
& Caen 1980). If this binding depends on multiple interactions
(Santoro & Cummingham 1979) these could also induce distortions
in the platelet membrane sufficient to activate phospholipase A_2
and thereby the prostaglandin cascade, resulting in the release
reaction as observed. This, therefore, represents yet another
way in which prostaglandins could participate in thrombogenesis.

Our knowledge of arterial thrombogenesis is still sufficiently fragmentary that the most reasonable way to conclude is with two quotations which counsel caution:
"Un point de vue unique est toujours faux" (Paul Valéry);   and
"For the belief in a single truth and in being the possessor thereof is the root cause of all evil in the world" (Max Born).

ACKNOWLEDGEMENTS

We wish to thank the Minna-James-Heineman Stiftung of Hanover for a Research Fellowship for Michael Kratzer; and the Fritz Thyssen Stiftung of Cologne and the British Heart Foundation for support.

REFERENCES

Begent, N.A., and Born, G.V.R., 1970, Growth rate in vivo of platelet thrombi, produced by iontophoresis of ADP, as a function of mean blood flow velocity, Nature (Lond.) 227: 926.

Born, G.V.R., 1977, Fluid-mechanical and biochemical interactions in haemostasis, Br.Med.Bull. 33(3): 193.

Born, G.V.R., 1979a, Arterial thrombosis and its prevention, In: Proc. VIII World Congress Cardiology, S. Hayase and S. Murao, eds., pp. 81-91, Excerpta Medica, Amsterdam.

Born, G.V.R., 1979b, Possible role for chlorpromazine in protection against myocardial infarction, The Lancet, April 14: 822.

Born, G.V.R., Bergqvist, D., and Arfors, K.E., 1976, Evidence for inhibition of platelet activation in blood by a drug effect of erythrocytes, Nature 259(5540): 233.

Born, G.V.R., and Richardson, P.D., 1980, Activation time of blood platelets, J.Membr.Biol. 57: 87.

Born, G.V.R., and Wehmeier, A., 1979, Inhibition of platelet thrombus formation by chlorpromazine acting to diminish haemodynamically induced haemolysis, Nature 282: 212.

Colantuoni, G., Hellums, J.D., Moake, J.L., and Alfrey, C.P. Jr., 1977, The response of human platelets to shear stress at short exposure times, Trans.Am.Soc.Artif.Intern.Organs 23: 626.

Constantinides, P., 1966, Placque fissures in human coronary thrombosis, J.Atheroscler.Res. 6: 1.

Davies, K.J., 1981, Pathological basis for deposition of platelets, Phil.Trans.Royal Soc. (in press).

Didisheim, P., Pavlovsky, M., and Kobayashi, I. 1972, Factors that influence or modify platelet function, Ann.NY Acad.Sci., H.J. Weiss, ed., p. 307.

Fox, A.L., and Hugh, A.E., 1966, Localisation of atheroma; a theory based on boundary layer separation, Br.Heart J. 28: 388.

Friedman, M., and Byers, S.O., 1965, Induction of thrombi upon pre-existing arterial placques, Amer.J.Pathol. 46: 567.

Gaarder, A., Jonsen, J., Laland, S., Hellem, A., and Owren, P.A., 1961, Adenosine diphosphate in red cells as a factor in the adhesiveness of human blood platelets, Nature 192: 531.

Gryglewski, R.J., Bunting, S., Moncada, S., Flower, R.J., and Vane, J.R., 1976, Arterial walls are protected against deposition of platelet thrombi by a substance (prostaglandin X) which they make from prostaglandin endoperoxides, Prostaglandins 12: 685.

Goldsmith, H.L., 1972, The flow of model particles and blood cells and its relation to thrombogenesis, In: Progress in haemostasis and thrombosis, T.H. Spaet, ed., Vol. 1, pp. 97-139, Grune and Stratton, New York.

Hellums, J.D., and Brown, C.H., 1977, Blood cell damage by mechanical forces, In: Cardiovascular flow dynamics and measurements, N.H. Hwang and N.A. Norman, eds., University Park Press, Baltimore, WG 106 NIII 1975.

Kratzer, M.A.A., and Born, G.V.R., 1981, Free ATP in blood during haemorrhage. VIII Int.Congr.Thromb.Haemostas. (Toronto, Canada, July 1981).

Legrand, Y.G., Karnignian, A., Le Francier, P., Fauvel, F., and Caen, J.P., 1980, Evidence that the collagen-derived nona-peptide is a specific inhibitor of platelet-collagen interaction, Biochem.Biophys.Res.Comm. 96: 1579.

Mackay, A.V.P., Healey, A.F., and Baker, J., 1974, The relationship of plasma chlorpromazine to its 7-hydroxy- and sulphoxide metabolites in a large population of chronic schizophrenics, Br.J.Clin.Pharmac. 1: 425.

Mills, D.C.B., and Roberts, G.C.K., 1967, Membrane active drugs and the aggregation of human blood platelets, Nature 213: 35.

Mustard, J.F., Murphy, E.A., Rowsell, H.A., and Downie, H.G., 1962, Factors influencing thrombus formation in vivo, Amer.J.Med. 33: 621.

Mustard, J.F., Moore, S., Packham, M.A., and Kinlough-Rothbone, R.L., 1977, Platelets, thrombosis and atherosclerosis, Prog. biochem.Pharmacol. 13: 312-325, Karger, Basel.

Odegard, O., 1967, Mortality on Norwegian psychiatric hospitals 1950-62, Acta genet., Basel 17: 137.

Packham, M.A., and Mustard, J.F., 1977, Clinical pharmacology of platelets, Blood 50(4): 555.

Richardson, P.D., 1973, Effect of blood flow velocity on growth rate of platelet thrombi, Nature (Lond.) 245: 107.

Richardson, P.D., Galletti, P.M., and Born, G.V.R., 1976, Regional administration of drugs to control thrombosis in artificial organs, Trans.Am.Soc.Artif.Intern.Organs 22: 22.

Rowntree, L.G., and Shionya, T., 1927, Studies in experimental extra-corporeal thrombosis. Part 1: Methods for the direct observation of extra-corporeal thrombus formation, J.Exp. Med. 46: 7.

Santoro, S.A., and Cummingham, L.W., 1979, Fibronectin and the multiple-interaction model for platelet-collagen adhesion, Proc.Nat.Acad.Sci. USA 76: 2644.

Schmid-Schönbein, H., Rohling-Winkel, I., Blasberg, P., Jungling, E., Wehmeier, A., Born, G.V.R., and Richardson, P.D., 1979, Release of ADP from erythrocytes under high shear stresses in tube flow, Thromb.Haemostas. (Abstracts, VII int. Congress Thromb. Haemostas.), Abstract 0835, 349.

Seeman, P., 1972, The membrane action of anaesthetics and tranquillizers, Pharmac.Rev. 24: 583.

Turitto, V.T., and Baumgartner, H.R., 1975, Platelet interaction with subendothelium in a perfusion system: physical role of red blood cells. Microvasc.Res. 9(3): 335.

Turitto, V.T., Benis, A.M., and Leonard, E.F., 1972, Platelet diffusion in slowing blood. Ind.Eng.Chem.Fund. 11: 216.

Wilner, G.D., Nossel, H.L., and LeRoy, E.C., 1968, Aggregation of platelets by collagen, J.Clin.Invest. 47: 2616.

Zahavi, J., and Schwartz, G., 1978, Chlorpromazine and platelet function, Lancet ii: 164.

PLATELET AND VASCULAR PROSTAGLANDINS IN URAEMIA, THROMBOTIC MICROANGIOPATHY AND PRE-ECLAMPSIA

G. de Gaetano[*], M. Livio[*], M.B. Donati[*] and G. Remuzzi

[*] Istituto di Ricerche Farmacologiche "Mario Negri",
Milan, and Division of Nephrology and Dialysis,
Ospedali Riuniti, Bergamo, Italy

The metabolism of arachidonic acid in platelets and vascular cells is often altered in clinical conditions associated with haemorrhagic or thromboembolic complications.

We have focused on the one hand in uraemia as a condition frequently complicated by bleeding episodes and, on the other, on thrombotic microangiopathy (thrombotic thrombocytopenic purpura,TTP; haemolytic uraemic syndrome,HUS) and pre-eclampsia as conditions characterized by uncontrolled intravascular platelet activation.

The observation that prostaglandin synthesis may be regulated by factors present in normal human plasma (Saeed et al.,1977; Mac Intyre et al.,1978) prompted us to investigate whether such plasmatic control was altered in the clinical situations mentioned.

Venous specimens removed from uraemic patients during the institution of an artero-venous shunt for haemodialysis generated significantly more prostacyclin (prostaglandin I_2,PGI_2) than control vessels (Remuzzi et al.,1977) . Similar findings were subsequently reported in aortic tissue from nephrectomized rats and arterial tissue from uraemic patients (Leithner et al.,1978). Uraemic plasma showed a greater capacity to stimulate PGI_2 synthesis by vascular rings or endothelial cultured cells (Remuzzi et al.,1978, Defreyn et al.,1980).

This suggests that altered prostacyclin generation in vessel wall from uraemic patients is mediated by plasma. In contrast both malondialdehyde (MDA) and thromboxane B_2 generated in response to relatively high concentrations of arachidonic acid or thrombin were significantly lower in platelets from uraemic patients than from controls. Uraemic plasma inhibited MDA generation in normal platelets, and normal plasma partially corrected the defect in uraemic platelets (Remuzzi et al.,1980a).

This suggests that an imbalance in the plasmatic regulation of prostaglandin metabolism in platelets and vessel wall from uraemic patients may contribute to their bleeding tendency.

Thrombotic microangiopathy is characterized by thrombocytopenia, haemolytic anemia, neurological abnormalities and/or renal failure. Microthrombi occluding arterioles and capillaries of different organs are found on pathological examination. This reflects the occurrence of widespread intravascular platelet aggregation, a crucial event in the pathogenetic sequence of thrombotic microangiopathy.

In three patients we studied with TTP or HUS (Remuzzi et al., 1978; Donati et al.,1980), no prostacyclin activity was released from vascular specimens obtained during the acute phase of the disease, suggesting that prostacyclin might be the physiological inhibitor of platelet aggregation, postulated as defective in a patient with TTP described by Byrnes and Khurana (1977). Plasma taken from all three patients on admission had very low capacity, if any, to stimulate vascular prostacyclin synthesis. Treatment with plasma exchange or infusion led to rapid clinical improvement and the patient's plasma recovered its capacity to stimulate prostacyclin generation. A deficiency of the plasma factor(s) stimulating vascular PGI_2 activity was therefore suggested as having some role in the pathogenetic sequence of thrombotic microangiopathy (Remuzzi et al.,1978). This "missing factor" hypothesis has gained further support from more recent observations that plasma levels of 6 keto $PGF_{1\alpha}$ (the chemically inactive derivative of PGI_2) were very low or undetectable in patients with TTP or HUS (Hensby et al., 1979; Machin et al.,1980; Webster et al.,1980). In a patient described recently (Remuzzi et al.,1980c), deficiency of the plasma factor persisted for at least one year after clinical remission from HUS without recurrence. A similar deficiency was detected in two of this patient's four offspring, who had never suffered microangiopathic episodes.

This suggests that — at least in some cases of HUS — the plasma defect might be genetically determined. Deficient PGI_2

stimulating activity in plasma would not normally result in any clinical sign of disturbed platelet function as long as no aetiological
agent such as endotoxin triggers the pathogenetic sequence of thrombotic microangiopathy (Donati et al.,1980).

Preliminary data in two patients studied during the acute
phase of HUS indicate that an increased tendency of activated platelets to generate thromboxane A_2 might be an additional factor fa-
vouring disseminated intravascular platelet aggregation in this
syndrome. Whether plasma modulates the exaggerated metabolism of
arachidonic acid in platelets from HUS patients has not yet been
clarified.

Pre-eclampsia is a major cause of morbidity and death for the
pregnant woman and her foetus. Signs of consumptive coagulopathy
frequently accompany hypertension, oedema and proteinuria, the
triad characteristic of this syndrome. Pathological examination
may show placental and glomerular vessels occluded by microthrombi,
and utero-placental ischemia appears to play a central role in the
pathogenesis.

We have recently reported that PGI_2 production is significantly depressed in umbilical and placental vessels from patients with
severe pre-eclampsia in comparison to normal pregnancy (Remuzzi et
al.,1980b). Reduced PGI_2 production has now be confirmed in umbilical artery (Dowing et al.,1980), in amniotic fluid (Bodzenta et
al.,1980) and in plasma (Bussolino et al.,1980). This maternal
hypertension, platelet consumption and reduced placental perfusion
could be triggered or maintained by a defect of the mechanism(s)
leading in normal pregnancy to increased levels of PGI_2. Indeed
in normal human pregnancy both foetal and maternal vessels produce
larger amounts of PGI_2 than vessels from non-pregnant women (Remuzzi et al.,1979; Lewis et al.,1980). This implies that an increase in vasodilatory PGI_2 may account for the low peripheral
resistance and the high renin activity of normal pregnancy.

This pathogenetic interpretation is reinforced by earlier
observations that pregnant rats fed a vitamin E-deficient fat diet
developed eclamptic crises (Stamler,1959). Indeed, vitamin E
deficiency has lately been reported to impair PGI_2 production in
rats (Okuma et al.,1980). Moreover administration of indomethacin
to sheep is followed by a marked increase in the resistance of uterine and placental vascular beds (Rankin et al.,1979).

Plasmatic regulation of vascular PGI_2 generation in pregnancy
has recently been studied (Remuzzi et al.,1981) . No significant
difference was found between non-pregnant and pregnant women during

early pregnancy, but a significant reduction of prostacyclin stimulating activity was observed in plasma during late normal pregnancy. In patients with severe pre-eclampsia this plasmatic activity
was within the range of control non-pregnant women, but significantly higher than comparable women with normal pregnancy.

These results are surprising and apparently difficult to
reconcile with the good correlation between high plasmatic activity
and high vascular PGI_2 in uraemic patients and low plasmatic activity and low vascular PGI_2 in patients with thrombotic microangiopathy. Possibly more than one mechanism operates in the control
of vascular prostacyclin production in normal pregnancy. Perhaps
the striking similarity between the behavior of "plasma factor"
and the response of blood pressure to angiotensin II in normal and
complicated pregnancy (Ferris,1978) offers a key to a better understanding of the role of prostacyclin and its regulation in pregnancy. It seems pertinent to mention here that in women with recurrent
spontaneous abortion a plasmatic activity (linked to the IgG fraction) inhibiting the release of PGI_2 from aortic rings has recently
been found (Carreras et al.,1980).

CONCLUSIONS

The nature of the plasma component(s) modulating platelet
and vascular prostaglandin synthesis in uraemia, thrombotic
microangiopathy and pregnancy is unknown. Whether such factor(s)
are identical to the endogenous modulator(s) described in normal
plasma is still to be clarified.

A crucial step in this phenomenon might lie in the balance
between free radical formation and removal in plasma (see for
detailed discussion Donati et al.,1980). An imbalance in the
synthesis of metabolites of endogenous arachidonic acid in some
clinical conditions such as those discussed in this paper is not
necessarily corrected either by the removal of the products generated in excess (achievable for instance using aspirin or more selective prostaglandin synthesis inhibitors) or by replacement of the
defective compound (for instance by infusion of prostacyclin or one
of its stable analogues). Indeed beneficial effects of intravenous
infusion of prostacyclin in TTP or HUS have been reported in some
patients (Webster et al.,1980) but not in others (Hensby et al.,
1979;Budd et al.,1980). On the other hand patients with pre-
eclampsia reportedly benefit from aspirin treatment at doses presumably inhibiting prostacyclin generation (Crandon and Isherwood,
1979).

Thus, although any new therapeutic attempt to improve the

natural course of these diseases must be encouraged, it would be premature to draw any pharmacological implications from pathogenetic hypotheses still awaiting full confirmation.

Acknowledgements

Work mentioned in this paper was performed with the support of the Associazione Bergamasca per lo Studio delle Malattie Renali and Italian National Research Council ("Farmacologia Clinica e Malattie Rare" and Gruppo Nazionale di Ricerca "Tecniche sostitutive di funzioni d'organo").

REFERENCES

Bodzenta, A., Thomson, J.M., and Poller, L., 1980, Prostacyclin activity in amniotic fluid in pre-eclampsia, Lancet ii: 650.
Budd, G.T., Bukowski, R.M., Lucas, F.V., Cato, A.E., and Cocchetto, D.M., 1980, Prostacyclin therapy of thrombotic thrombocytopenic purpura, Lancet ii: 915.
Bussolino, F., Benedetto, C., Massobrio, M., and Camussi, G., 1980, Maternal vascular prostacyclin activity in pre-eclampsia, Lancet ii: 702.
Byrnes, J.J., and Khurana, M., 1977, Treatment of thrombotic thrombocytopenic purpura with plasma, N.Eng.J.Med. 297: 1386.
Carreras, L.D., Defreyn, G., Machin, S.J., and Vermylen, J., 1980, Inhibition of prostacyclin release from rat aorta by IgG from a patient with lupus anticoagulant and recurrent thrombosis, In: Abstracts Sixth International Congress on Thrombosis, Montecarlo, October 1980, n. 187.
Crandon, A.J., and Isherwood, D.M., 1979, Effect of aspirin on incidence of pre-eclampsia, Lancet i: 1356.
Defreyn, G., Vergara Dauden, M., Machin, S.J., and Vermylen, J., 1980, A plasma factor in uraemia which stimulates prostacyclin release from cultured endothelial cells, Thromb. Res. 19: 695.
Donati, M.B., Misiani, R., Marchesi, D., Livio, M., Mecca, G., Remuzzi G., and de Gaetano, G., 1980, Hemolytic-uremic syndrome, prostaglandins, and plasma factors, In: Hemostasis, Prostaglandins and Renal Disease, G. Remuzzi, G. Mecca, and G. de Gaetano, eds., pp. 283-290, Raven Press, New York.
Downing, I., Shepherd, G.L., and Lewis, P.J., 1980, Reduced prostacyclin production in pre-eclampsia, Lancet ii: 1374
Ferris, T.F., 1978, Postpartum renal insufficiency, Kidney Int. 14: 383.
Hensby, C.N., Lewis, P.J., Hilgard, P., Mufti, G.J., Hows, J., and Webster, J., 1979, Prostacyclin deficiency in thrombotic thrombocytopenic purpura, Lancet ii: 748.

Leithner, C., Winter, M., Silberbauer, K., Wagner, O., Pinggera, W., and Sinzinger, H., 1978, Enhanced prostacyclin availability of blood vessels in uraemic humans and rats, In: Dialysis, Transplantation, Nephrology, B.H.B. Robinson and J.B. Hawkins, eds., pp.418-422, Pitman Medical, Tunbridge Wells, Kent, U.K.

Lewis, P.J., Boylan, P., Friedman, L.A., Hensby, C.N., and Drowning, I., 1980, Prostacyclin in pregnancy, Br.Med.J. 280: 1581.

Machin, S.J., Defreyn, G., Chamone, D.A.F., and Vermylen, J., 1980, Plasma 6-keto-PGF$_{1\alpha}$ levels after plasma exchange in thrombotic thrombocytopenic purpura, Lancet i: 661.

MacIntyre, D.E., Pearson, J.D., and Gordon, J.L., 1978, Localisation and stimulation of prostacyclin production in vascular cells, Nature 271: 549.

Okuma, M., Takayama, H., and Uchino, H., 1980, Generation of prostacyclin-like substance and lipid peroxidation in vitamin E-deficient rats, Prostaglandins 19:527.

Rankin, J.H.G., Berssenbrugge, A., Anderson, D., and Phernetton, T., 1979, Ovine placental vascular response to indomethacin, Am. J. Physiol. 236: H61-H64.

Remuzzi, G., Cavenaghi, A.E., Mecca, G., Donati, M.B., and de Gaetano, G., 1977, Prostacyclin-like activity and bleeding in renal failure, Lancet ii: 1195.

Remuzzi, G., Marchesi, D., Livio, M., Schieppati, A., Mecca G., Donati, M.B., and de Gaetano, G., 1980a, Prostaglandins, plasma factors and hemostasis in uraemia, In: Hemostasis, Prostaglandins and Renal Disease, G. Remuzzi, G. Mecca, and G. de Gaetano, eds. pp. 273-281, Raven Press, New York.

Remuzzi, G., Marchesi, D., Zoja, C., Muratore, D., Mecca, G., Misiani, R., Rossi, E., Barbato, M., Capetta, P., Donati, M.B., and de Gaetano, G., 1980b, Reduced umbilical and placental vascular prostacyclin in severe pre-eclampsia, Prostaglandins 20: 105

Remuzzi, G., Mecca, G., Livio, M., de Gaetano, G., Donati, M.B. Pearson, J.D., and Gordon, J.L., 1980, Prostacyclin generation by cultured endothelial cells in haemolytic uraemic syndrome, Lancet i: 656.

Remuzzi, G., Misiani, R., Marchesi, D., Livio, M., Mecca, G., de Gaetano, G., and Donati, M.B., 1978, Haemolytic-uraemic syndrome: Deficiency of plasma factor(s) regulating prostacyclin activity?, Lancet ii: 871.

Remuzzi, G., Misiani, R., Muratore, D., Marchesi, D., Livio, M., Schieppati, A., Mecca, G., de Gaetano, G., and Donati, M.B., 1979, Prostacyclin and human foetal circulation, Prostaglandins 18: 341.

Remuzzi, G., Zoja, C., Marchesi, D., Schieppati, A., Mecca, G., Misiani, R., Donati, M.B., and de Gaetano, G., 1981, Plasmatic regualtion of vascular prostacyclin in pregnancy, Br.Med.J. (in press).

Saeed, S.A., McDonald-Gibson, W.J., Cuthbert, J., Copas, J.L.,
    Schneider C., Gardiner, P.J., Butt, N.M., and Collier, H.O.J.
    1977, Endogenous inhibitor of prostaglandin synthetase,
    Nature 270: 32.
Stamler, F.W., 1959, Fatal eclamptic disease of pregnant rats fed
    anti-vitamin E stress diet, Am.J.Pathol. 35: 1207.
Webster, J., Rees, A.J., Lewis, P.J., and Hensby, C.N., 1980,
    Prostacyclin deficiency in haemolytic-uraemic syndrome,
    Br.Med.J. 281: 271.

PHARMACOLOGICAL AND DIETARY CONTROL OF PLATELET VESSEL WALL INTER-

ACTIONS IN NORMAL AND PATHOLOGICAL CONDITIONS

E. Tremoli, C. Galli and R. Paoletti

Institute of Pharmacology and Pharmacognosy, Universi-

ty of Milan, Via A. Del Sarto 21, 20129 Milan, Italy

A number of recent experimental and clinical studies deals with the role of platelets in the development of atherosclerosis and of thromboembolic diseases (1,2). The interaction between platelets and vessel walls has been the object of extensive studies on the basis of the recent discovery of the modulators of platelet homeostasis, prostacyclin and thromboxane (3,4).

The relationship between platelets and prostaglandins and the effect of antiaggregating compounds and/or different dietary treatments in animal studies and in human normal and pathological conditions will be discussed.

1) AGGREGATING AGENTS AND PLATELET RESPONSES

Most of our informations about the mechanisms involved in platelet aggregation originates from in vitro data using differ-ent platelet functional tests (5). Although the physiological relevance of such data has been questioned (6), the use of these functional tests has been helpful in elucidating aspects of platelet physiology and in evaluating the mechanism of action of antiplatelet drugs (7). Many substances present in vivo, i.e. collagen, thrombin and arachidonic acid induce platelet aggrega-tion and release reaction (8). Platelets, however, during aggrega-tion and release can synthetize active compounds which in turn sustain the aggregatory process (9). In addition many of the inducers may act through independent mechanisms, not all as yet completely clarified (10).

Thrombin, for instance, induces platelet aggregation through at least three mechanisms: the release of ADP from dense bodies, the activation of the arachidonic acid pathway and the formation of a platelet aggregating factor (PAF) (10,11,12). The same is true for the collagen induced aggregation (13).

The first in vitro response of platelets to all aggregating agents, excluding epinephrine (16), is the change in shape (14). Then depending upon the concentrations of the inducers considered, reversible or irreversible aggregation occurs. The reversible or irreversible response of platelet aggregation is strictly dependent on the strenght of the inducer, as shown by Holmsen (9), which is defined as the ability of a single agent to evoke the sequence of responses (shape change - aggregation - release from dense bodies - arachidonic acid activation - release from alpha granules).

Table 1 shows some of the best well known aggregating agents and the comparative strenght of the evoked responses. Thrombin and collagen stimulate platelets and induce all the sequence of responses, including the formation of PAF.

Table 1. Aggregating agents and platelet response.

Inducer	Shape Change	Aggre-gation	AA oxi-dation	Release from dense bodies	Release from alpha granules
Thrombin	+	+	+	+	+
Collagen	+	+	+	+	+
PGG_2/PGH_2	+	+	+	+	+
ADP	+	+	$\pm$	+	?
Epinephrine	-	+	$\pm$	+	?

Modified from Holmsen (1979).

The same is true for cyclic endoperoxides with the exception of PAF generation. The ADP and epinephrine induced aggregation involves in a lesser extent the arachidonic acid metabolism, although acetylsalycilic acid is able to inhibit both the ADP and epinephrine second wave aggregation (15), corresponding to the release reaction. This reaction consists in the release of material contained in three type of granules: dense bodies, alpha granules and lysosomal organelles (16,17,18). In particular the non metabolic pool of adenine nucleotides, 5 hydroxytriptamine (5 HT), calcium ions, pyrophosphate and antiplasmin are stored in the dense bodies (19,20). The alpha granules contain fibrinogen, beta- thromboglobulin (beta-TG) (21), platelet factor 4 (PF4) (22,23), the platelet derived growth factor (PDGF) (24) and the vascular permeability growth factor (25). Finally the lysosomal organelles contain mainly hydrolases (26).

During platelet aggregation the coagulation pathways are also activated at various stages (27). Platelets possess a number of coagulant activities leading to fibrin formation (28). The platelet membrane, during shape change, following activation becomes a catalytic surface which promotes the formation of thrombin in the presence of factor Xa, factor V and calcium ions. This activity is defined as platelet factor 3 (PF3) (29). Other activities described by Walsh can cooperate, in particular conditions, to the activation of the coagulation pathway (30). Recently it has been shown that platelets as cancer cells can directly activate factor X, suggesting the existence of a cellular pathway of blood coagulation (31,32,33). Thus platelets during activation can be the site of local generation of thrombin, which in turn can stimulate other platelets to aggregate, resulting in the formation of the platelet thrombus (Fig. 1).

2) PROSTAGLANDINS FORMATION AND EFFECTS ON PLATELETS

Prostaglandins and thromboxane are produced from 20 carbon polyunsaturated fatty acids, such as dihomogammalinolenic acid (DHGLA), arachidonic acid (AA) and eicosapentaenoic acid (EPA) respectively precursors of the 1,2 and 3 series (34,35,36). The prostaglandins and thromboxanes of the series 2, all derived from arachidonic acid, have been shown to have the greatest physiological importance (Table 2).

Smith and Willis firstly provided evidence that during aggregation induced by thrombin, platelets produce prostaglandins, i.e. the stable prostaglandins PGE_2 and $PGF_{2\alpha}$ (37). The linkage between platelet aggregation and the prostaglandin production was not immediately clear, since it had already shown that these two prostaglandins (PGs) were practically ineffective in inducing platelet aggregation (38). The observation by Silver that

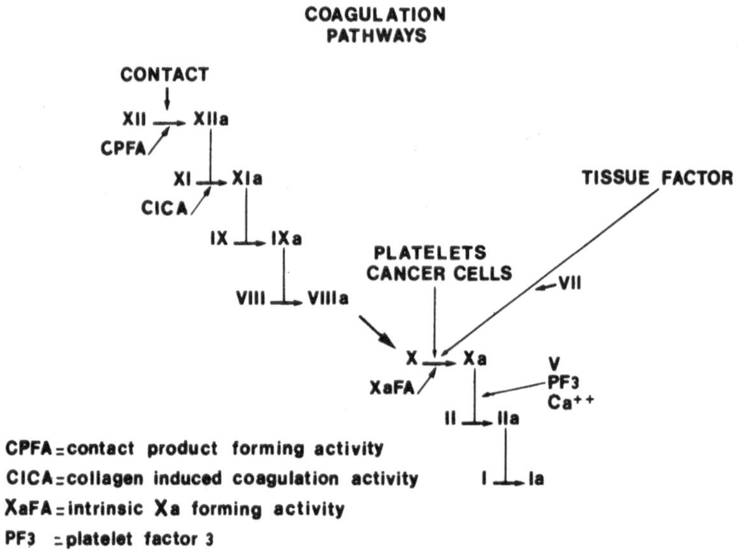

COAGULATION
PATHWAYS

CONTACT

XII → XIIa

CPFA

XI → XIa

CICA

IX → IXa

PLATELETS
CANCER CELLS

TISSUE FACTOR

VIII → VIIIa

← VII

X → Xa

XaFA

V
PF3
Ca++

II → IIa

CPFA = contact product forming activity

CICA = collagen induced coagulation activity

XaFA = intrinsic Xa forming activity

PF3 = platelet factor 3

I → Ia

Fig. 1 – Platelet coagulant activities.

Table 2. Polyunsaturated fatty acids precursor of prostaglandins
and thromboxanes.

DIHOMOGAMMALINOLENIC ACID   prostaglandins series 1

20:3 (n-6)

ARACHIDONIC ACID   prostaglandins series 2

20:4 (n-6)

EICOSAPENTAENOIC ACID   prostaglandins series 3

20:5 (n-3)

---

arachidonic acid, the 20 carbon fatty acid precursor of the PGs
series 2, was able to induce platelet aggregation (39), was
immediately followed by the discovery of cyclic endoperoxides
(PGG_2 and PGH_2) (40) intermediates, with aggregatory properties.

The structure of thromboxane A_2 (TXA_2), the most potent platelet aggregating substance so far discovered, and of thromboxane B_2 (TXB_2), a more stable metabolite, was then established by Samuelsson (41).

The synthesis of PGs and TXs is low in basal conditions, increasing when platelets are stimulated. The AA released from the membrane phospholipids, following activation of phospholipases (42), is a substrate for lipooxygenase and cyclooxygenase.

The lipooxygenase pathway of AA in platelets leads to the formation of hydroperoxy and hydroxy fatty acids, i.e. HPETE and HETE, which possess chemotactic activity (43). Other cells, such as polymorphonuclear leucocytes have been recently shown to synthetize the biologically active leukotrienes, also lipooxygenase products, which are important in the anaphylactic processes (44,45). No informations are available on the possible interactions between these products of leucocytes and platelet behavior. On the other hand the activation of cyclooxygenase pathway in platelets leads to the formation of the stable PGs, E_2, $F_2\alpha$, and D_2 and of the thromboxane A_2 and B_2. The products of AA metabolism synthetized by platelets are shown in Fig. 2. Since aspirin and indomethacin have been demonstrated to inhibit platelet aggregation and platelet PGs production, it has been suggested that platelet aggregation was associated with PGs production (46). The use of the PGE_2 as aggregating agents and $PGF_{2\alpha}$ did not support this hypothesis. In contrast PGD_2 was found to be a potent inhibitor of platelet aggregation (47). The discovery of cyclic endoperoxides and the identification of TXA_2, with potent aggregatory properties was followed by the demonstration that these AA derivatives are strong inducers of platelet shape change and release reaction, indicating that these compounds are directly involved in platelet function (48,49).

An important link between platelets and arterial walls has been shown by the observation that prostacyclin (PGI_2), an AA metabolite not synthetized by platelets, but by the vessel walls, particularly by the endothelium (50,51) is the most powerful antiaggregating compound ever described (52,53). The relations between prostacyclin and thromboxane formation and the effects on platelets is, therefore, considered the basis for an important homeostatic mechanism regulating the interactions between platelets and vessel walls under normal and pathological conditions.

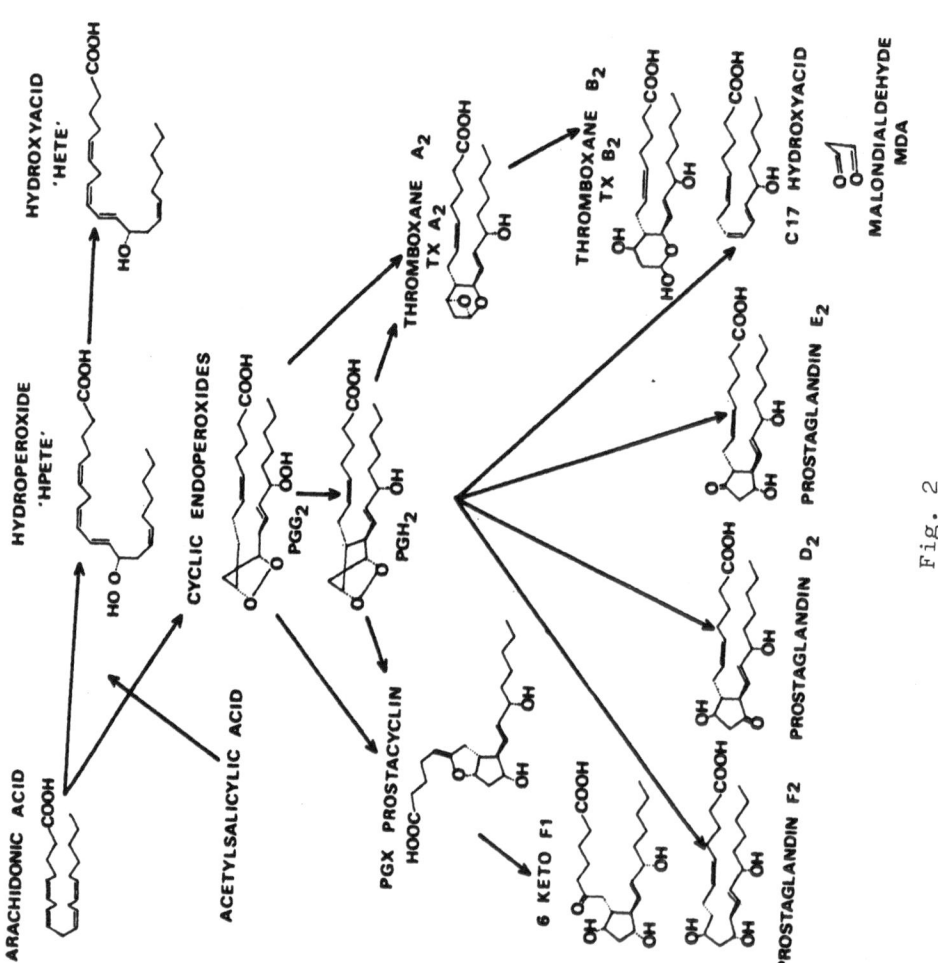

Fig. 2

3) PLATELET AGGREGATION AND PLATELET CYCLIC AMP LEVELS

Agents inhibiting platelet aggregation and the release reaction increase the intracellular levels of cyclic AMP (cAMP) either by stimulating the adenylcyclase (54) or inhibiting cyclic phosphodiesterase (55). This finding indicates that cAMP levels are related to platelet response to aggregating agents (56). It has been postulated that agents triggering platelet aggregation do this through a decrease of cAMP levels in platelets (57). Epinephrine, collagen and thrombin, in fact, have been shown to induce a decrease of the basal cAMP levels (58). On the other hand this effect is suppressed by acetylsalycilic acid (59). On these basis it has been suggested that prostaglandins are involved in platelet function as cAMP mediators. On the contrary little informations are available on the possible role of cyclic GMP (60).

PGE_1, PGD_2 and PGI_2 have been shown to stimulate adenylcyclase, increasing cAMP levels in cell membranes (61,62). PGI_2 for instance possesses a strong activity on cAMP level stimulation, through activation of adenylate cyclase as compared to PGE_1. The kinetic of cAMP stimulation by PGI_2 and PGE_1 appears to be completely different in terms of time course, suggesting that different mechanisms are involved (63). Finally the existence of specific receptors for prostaglandins and thromboxane A_2 on platelet membrane has been demonstrated (64,65). Indirect observations (66,67) indicate that PGE_1 and PGI_2 act through the same receptor and PGD_2 possesses a different receptor site on the platelet membrane.

4) PLATELET HYPERSENSITIVITY TO AGGREGATING AGENTS ASSOCIATED TO ENHANCED PROSTAGLANDIN BIOSYNTHESIS

Increased platelet sensitivity and platelet prostaglandin biosynthesis are frequently associated with systemic diseases (68,69). Platelets from diabetic subjects for instance are more sensitive to in vitro aggregation induced by various agents and spontaneous platelet aggregation has been reported in this disease (70). Halushka et al. (71) have found that platelets from diabetic subjects synthetize higher amounts of prostaglandin E like substance, following stimulation with ADP, epinephrine, collagen and arachidonic acid.

Similar findings have been reported in patients suffering

from arterial thrombosis, transient ischaemic attacks and deep venous thrombosis (72,73). In addition individuals with familial history of hypertension, angina and myocardial infarction or stroke showed increased synthesis of prostaglandins and thromboxanes, suggesting a linkage between the tendency to develop cardiovascular disease and the platelet abnormality.

Patients with type IIa hypercholesterolemia, according to WHO (74), are known to be at increased risk of thrombosis and atherosclerosis (75,76). A platelet hypersensitivity to aggregating agents has also been described in this disease by several authors (77,78,79). The platelet hypersensitivity has also been associated with an enhanced production of malondialdehyde and thromboxane B_2 by platelets as shown in our laboratory (79,80) and confirmed by other groups. The observation of an altered lipid peroxidation and an increased formation of arachidonic acid metabolites in type IIa hypercholesterolemia observed in human is confirmed also by animal studies (81).

Cholesterol "per se" could be responsible of these abnormalities. Cholesterol enrichment of platelets results in the enhanced sensitivity to aggregating stimuli (82) in particular to epinephrine. Conflicting results however have been reported on the C/PL ratio in platelets from type IIa hypercholesterolemic patients. More recently, Stuart and coworkers showed that normal platelets incubated with cholesterol rich liposomes release higher amounts of arachidonic acid, which is then converted to thromboxane, following stimulation with thrombin (83).

Although the mechanisms responsible for the alterations of platelet response in type IIa hypercholesterolemia subjects are yet to be clarified, the high frequency of thrombotic episodes coupled with the observation of an altered platelet behaviour suggests that treatments aimed to the reduction of platelet aggregation and thromboxane synthesis and/or plasma LDL and total cholesterol levels are indicated in these patients.

The effects of Indobufen therapy at daily dose of 200 mg b.i.d. and those of a placebo were compared in a selected group of hypercholesterolemic subjects whose platelets showed a higher sensitivity to epinephrine and arachidonic acid induced aggregation and in a group of normolipidaemic subjects. The Indobufen administration completely suppressed epinephrine and arachidonic acid induced aggregation both in patients and in controls when aggregation was induced using threshold concentrations of aggregating agents. This effect lasted up to twelve hours from the last

Indobufen administration. The same was true for malondialdehyde
production by platelets stimulated with serial concentrations of
thrombin (range 2.5–25 U/ml), indicating that Indobufen therapy
resulted in the total inhibition of the capacity of platelets to
produce AA metabolites at least for a period of twelve hours.
When platelets from both controls and patients, however, have
been challenged with higher concentrations of epinephrine and AA,
in order to find the concentrations of aggregating agents reversing the inhibitory effect exerted by the drug, significantly
lower concentrations were found for platelets from type IIa
patients as compared with controls (Fig. 3). Thus the inhibitory
effect exerted by the antiaggregating compounds seems to be, at
least in type IIa patients, different from the one of normolipidaemic subjects, for platelet aggregation response.

The different response of platelets to the inhibitory effect
of Indobufen suggests that the potency of antiaggregating compounds should be checked also in pathological conditions, in
order to assess the most suitable therapeutic schedule.

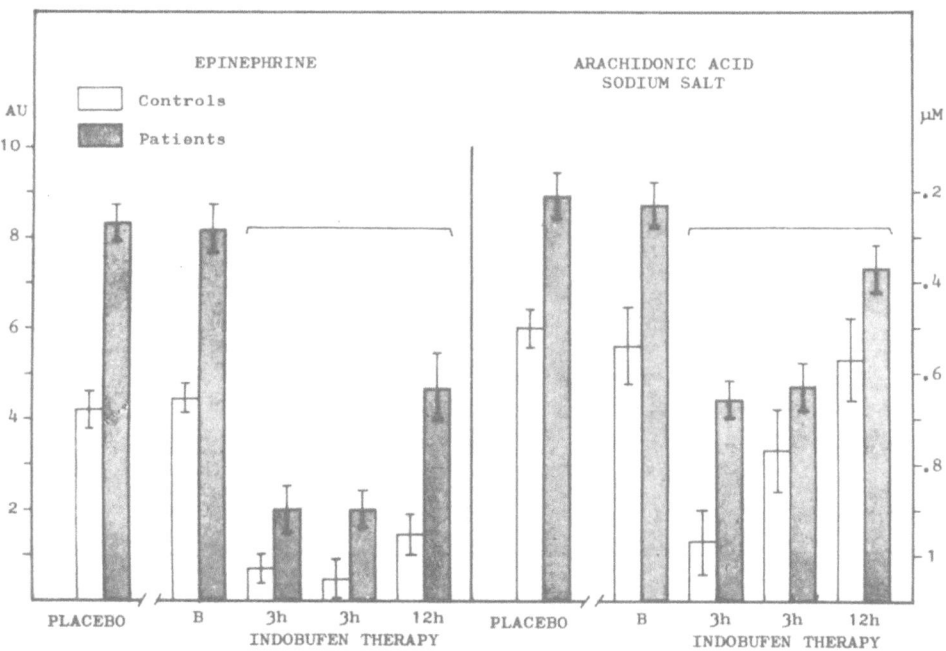

Fig. 3 - Overcoming concentrations (OCs) of Epinephrine and Ara-
         chidonic acid sodium salt during Indobufen administra-
         tion to normo and hypercholesterolemic subjects.

5) MODE OF ACTION OF ANTIAGGREGATING AGENTS

The elucidation of some of the mechanisms involved in platelet aggregation and in the release reaction has led to a better
understanding of the mode of action of specific agents interacting in vitro and in vivo with platelets at different steps of
their activation. Table 3 indicates some of the compounds exerting antiaggregating activity. The inhibitors of AA oxidation,
such as acetylsalicilic acid (ASA), indomethacin and other non
steroidal antiinflammatory drugs (NSAIDs), inhibiting platelet
cyclooxygenase (84), have been widely used for the control of
platelet aggregation (85). The evidence that these drugs are able
to inhibit also the vessel wall cyclooxygenase, interfering with
the synthesis of prostacyclin, has cast some doubts on the
clinical significance of these compounds (86). Current evidence
indicates that prostacyclin is largely responsible of the non
thrombogenicity of normal vascular surfaces (87). Platelet cyclooxygenase is, however, uniquely sensitive to ASA, because platelets, non nucleated cells, are unable to resynthetize the enzyme
irreversibly blocked by ASA and, in contrast, the endothelial
cell cyclooxygenase, which may be resynthetized, is less sensitive to the inhibitory effect of ASA (88,89). Therefore, several
studies have been designed in attempt to define the minimal ASA
dose inhibiting platelet aggregation and prostaglandin formation,
without affecting prostacyclin synthesis by the vessel walls.
Patrono and coworkers reported that low ASA doses, 10-30 mg/kg,
are sufficient to inhibit platelet TXB_2 production (90). No ASA
dose, however, which completely inhibited TXB_2 production by
platelets, without any inhibitory effect on PGI_2 production by
arterial and venous tissues has been found (91,92).

The choice of a treatment schedule with antiplatelet drugs
is also complicated by the fact that platelets from normal subjects and patients, show different sensitivity to the antiaggregatory agents and even to prostacyclin (93,94). In addition, following ASA administration the effect of two different drugs acetylsalicilic acid and salicilic acid should be considered (95). Salicilic acid, which is easily formed by plasma deacetylating enzymes (96), does not affect platelet aggregation, but is able to
prevent and reverse the inhibitory effect on platelet and vessel
wall cyclooxygenase exerted by ASA (95).

The phosphodiesterase inhibitors, such as dipyridamole and
other pyrimido-pyrimidine compounds inhibit at high concentra-

Table 3. Drugs interfering with platelet aggregation.

Activators of adenylate cyclase

Prostacyclin (PGI$_2$)

Prostaglandin E$_1$

Prostaglandin D$_2$

Inhibitors of Arachidonic acid oxidation

5-8-11-14 Eicosatetrayonate

Acetylsalicylic acid, Indobufen and other NSAIDS

Imidazol, 13-azaprostanoic acid

Vitamin E and other antioxidants

Inhibitors of cyclic phosphodiesterase

Papaverine

Pyrimido-Pyrimidine

Methylxantines

Other compounds

a) Antagonist of the inducers:

   ATP, Phentolamine, Dihydroergotamine

b) Inhibitors of the transmitter release:

   Chlorpromazine, Imipramine, Tetra-
   caine, Dibucaine

c) Thrombin inhibitors:

   Benzamidine derivatives, GYKI 14,451

tions the in vitro platelet aggregation, by enhancing cAMP levels
(97,98). In addition they potentiate the antiaggregating activity
of endogenous prostacyclin, thus enhancing the antithrombogenic
activity of the vessel walls (99). Although the antiaggregating
properties of dipyridamole are difficult to be observed using
physiological concentrations in the classical in vitro aggregating tests, dipyridamole has been shown to prolong the shortened
platelet survival time both in animals and in humans (100,101).

Among the adenylcyclase activators synthetic prostacyclin
and its stable analogs, such as carboprostacyclin are now considered of potential usefulness in the therapy of thromboembolic diseases (102).

Much attention, on the other hand, is now focused on the
development of compounds interfering with specific enzymes and/or
receptors. In particular molecules which inhibit selectively
platelet thromboxane synthesis, such as imidazole (103) and or
the thromboxane A_2 platelet receptor, such as 13-azaprostanoic
acid are now studied extensively (64). Finally compounds which
mimick or potentiate the release of prostacyclin, such as suloc-
tidil (104,105) have been shown to possess a consistent antithrombotic activity.

An important question, however, to be answered is the clinical significance in thromboembolic diseases of the antiaggregatory agents.

5a) ANTIPLATELET DRUGS IN CEREBROVASCULAR DISEASES AND ISCHAEMIC
    HEART DISEASE

The hypothesis that most transient ischaemic attacks, particularly in the carotid artery territory, have a thromboembolic
basis, supports the concept that drugs which interfere with platelet aggregation could be useful in this pathological condition.
Among the compounds known to possess antiaggregatory activity,
ASA and NSAIDs have been widely employed in several trials
designed to determine the possible clinical benefit in ischaemic
cerebrovascular disease.

ASA, on the basis of the overall studies, has been shown to
possess a protective effect in males, with respect to reduced
morbidity and mortality from cerebral infarction (106) and in
both sexes in the respect to the reduction in carotid transient
ischaemic attacks (106).

Sulfinpyrazone in contrast used alone or in combination with

ASA has not been shown to be more effective of ASA alone (107). As far as survival from myocardial infarction, the results obtained in the Aspirin Myocardial Infarction Study (AMIS) (108), are negative in the sense of a protective effect of aspirin on the incidence of total mortality during the three year follow up, although a slight but not statistically significant decrease in the percentage of definite non fatal myocardial infarctions has been shown in the aspirin group. The rationale of the AMIS study, however, was based on the data emerging from the Elwood study (109) and the Coronary Drug Project (110) in 1975. Thus the study was designed using 1 gr. ASA daily, which at the present time, is considered a high dose of aspirin.

A multicentre double blind clinical study has been concluded in 1979, using sulfinpyrazone (200 mg four times daily) and placebo. These two treatments were compared on the rates of cardiac mortality in 1475 survivors of myocardial infarction. The results of this clinical study indicate that sulfinpyrazone reduces the incidence of sudden cardiac death, at least during the first six months of follow up (111). These results are however challenged by the FDA on the basis of the patient selection.

Finally the effects of the combination of dipyridamole with ASA compared with the effects of ASA alone and with placebo on mortality and morbidity in patients with myocardial infarction have been investigated in a multicentre trial recently concluded (The Persantin-Aspirin Reinfarction Study-PARIS) (112). Although at the end of the total follow up the total mortality and morbidity were lower but not statistically significant in the drug treated groups, the analysis of the data has shown some important points. Firstly, the life-table rates for coronary mortality and coronary incidence were about 50% and 30% lowered in dipyridamole/ASA and ASA groups as compared with placebo from 8 to 24 months of follow up. In addition patients admitted to the study within 6 months from the myocardial infarction showed the largest reductions in the total and coronary mortality both in the dipyridamole/ASA and in ASA groups, compared with placebo. Again the doses of ASA are too high for a selective effect on platelet cyclooxygenase.

THE DIETARY APPROACH

Several studies have shown that changes in dietary fatty acids not only modify plasma lipid levels, but also affect the thrombotic tendency in experimental animals and platelet aggrega-

tion in humans. It is, in fact, generally accepted that the type
and the amount of dietary fat can influence the genesis and the
course of atherosclerosis (113,144). Since atherosclerosis and
thrombosis are related phenomena it is interesting that dietary
fats play also a role in the haemostatic system. Dietary saturated fatty acids, with more than 14 carbons such as palmitic and
stearic acid have been shown to enhance thrombus formation in
experimental animals, whereas oleic acid is neutral (115). Much
interest is now focused on the mechanisms underlying the effects
of dietary lipids on platelet function based on possible changes
of platelet prostaglandin biosynthesis. In particular the type of
dietary fat can influence the fatty acid composition of tissue
lipids, including platelet lipids, and especially the PG precursor fatty acid pool and/or the activity of the biosynthetic
enzymes (116). For instance it has been reported that tissue
prostaglandins are enhanced following administration of diets
rich in linoleic acid (117) possibly due to accumulation in
tissue phospholipids of the PG precursor arachidonic acid under
these dietary conditions. An excess of dietary linoleic acid
however may not necessarily produce an increase of arachidonic
acid in all tissues. In fact platelet phospholipids of animals
fed diets enriched with linoleic acid contain higher levels of
this fatty acid and lower levels of the major product of its
elongation and desaturation (arachidonic acid). In these experimental conditions a reduced response of platelets to thrombin and
to arachidonic acid induced aggregation were reported (118).
Enhanced formation of TXB_2 from exogenous arachidonic acid by
platelets of animals fed linoleic acid enriched diet (119) was
also observed, suggesting that the cyclooxygenase activity of
platelets is in some way depressed under these conditions.

The overall effects of diets with different linoleic acid
content in the production of PGI_2 or of PGI_2-like material by
arterial walls have been studied in various laboratories with
conflicting results. Some of the discrepancies may be attributed
to methodological difficulties in the determination and to problems in the selection of experimental design for "in vitro" studies. For instance the release of PGI_2 during incubation of manipu-
lated vascular tissue is not representative of a physiological
process giving an indication of the total biosynthetic capacity
of the tissue.

The recent observation that lipids of certain edible fish
are rich in eicosapentaenoic acid (EPA) has focused the attention

of investigators on the effects of this fatty acid on haemostatic parameters. Incubation of platelets with EPA does not stimulate aggregation, and inhibits aggregation induced by AA (120). Also platelets of animals and humans (Eskimos), fed EPA-rich diets are less sensitive to aggregating agents. There is, however no agreement about the mechanisms underlying these phenomena (121). It has been suggested, for instance, that from EPA a less active TXA_2 (122), or an antiaggregating product such as PGD_3 are formed (123). Alternative interpretations are inhibition of the platelet receptor for TXA_2 (36) or inhibition of the cycloxygenase for TXA_2 formation (124). Recently it has been shown that EPA is not a very good substrate for the cyclooxygenase, whereas is actively converted through the lipooxygenase (125). Also administration of EPA to experimental animals, reduces the production of PGI-like material by incubated aortas, suggesting that both the formation of pre-aggregatory and of antiaggregatory compounds is depressed by EPA. On the light of these observations it seems an over semplification to recommend the consumption of high levels of EPA as preventive or possibly therapeutic approach in thromboembolic disease.

In conclusion, manipulation of dietary lipids significantly affect platelet function and biochemistry thus providing the tool for dietary intervention in prethrombotic states. However, the limited knowledge of the biochemical processes underlying platelet aggregation and platelet-vessel walls interactions and of the effects of dietary changes on the single parameters modulating the overall process (e.g. TXA_2 and PGI_2 formation, platelet responsiveness to PGI_2) does not allow an immediate prediction of the consequence of the dietary approach.

REFERENCES

1. M.B. Stemerman, Haemostasis, thrombosis and atherogenesis, in "Atherosclerosis Reviews", vol. 6, M. Gotto and R. Paoletti,eds., Raven Press, New York (1979).
2. T.H. Spaet, Hemostatic homeostasis, Blood, 28:112 (1966).
3. M. Hamberg, J. Svensson and B. Samuelsson, Thromboxanes a new group of biologically active compounds derived from prostaglandin endoperoxides, Proc. Natl. Acad. Sci. USA, 72:2994 (1975).
4. S. Moncada, E.A. Higgs and J.R. Vane, Human arterial and venous tissues generate prostacyclin (prostaglandin X) a potent inhibitor of platelet aggregation, Lancet, 1:18

(1977).

5. J.F. Mustard and M.A. Packham, Factors affecting platelet function: adhesion, release and aggregation, Pharmacol. Rev. 22:97 (1970).

6. J.F. Mustard, W.D. Perry, R.L. Kinlough Rathbone, and M.A. Packham, Factors responsible for ADP induced release reaction of human platelets, Am. J. Physiol., 228:1757 (1975).

7. H.J. Weiss, The role of biologically active lipids in platelet aggregation, New Engl. J. Med., 293:531, 580 (1975).

8. B.B. Vargaftig, M. Chignard, and J. Lefort, Platelet secretion (release reaction). Mechanisms and pharmacology, in "Prostaglandins Immunopharmacology, Proceedings of the VII Int. Congress on Pharmacology", B.B. Vargaftig ed., Pergamon Press Publ., New York, p. 97 (1979).

9. H. Holmsen, Prostaglandin-Endoperoxide-Thromboxane synthesis and dense granule secretion as positive feed back loops in the preparation of platelet responses during the "basic release reaction", in "Prostaglandins Immunopharmacology, Proceedings of the VII ·Int. Congress on Pharmacology", B.B. Vargaftig ed., Pergamon Press, New York (1979).

10. R.L. Kinlough Rathbone, M.A. Packham, H.J. Reimers, J.P. Cazenave, and J.F. Mustard, Mechanisms of platelet shape change, aggregation and release induced by collagen, thrombin or A23, 187, J. Lab. Clin. Med., 90:707 (1977).

11. D.M. Tollefsen, J.R. Feagler, and P.W. Majerus, The binding of thrombin to the surface of human platelets, J. Biol. Chem., 249:2646 (1974).

12. M.A. Packham, M.A. Guccione, J.P. Greenberg, R.L. Kinlough-Rathbone, and J.F. Mustard, Release of ^{14}C serotonin during intra platelet changes induced by thrombin and collagen, Blood, 50:915 (1977).

13. M. Chignard, J.P. Le Couedic, M. Tence, B.B. Vargaftig, and J. Benveniste, The role of platelet activating factor in platelet aggregation, Nature, 279:799 (1979).

14. D.C.B. Mills and G.C.K. Roberts, Effects of adrenaline on human blood platelets, J. Physiol., 193:443 (1967).

15. M.G. Davey and E.F. Luscher, Actions of thrombin and other coagulant and proteolytic enzymes on blood platelets, Nature, 216:857 (1967).

16. H. Holmsen, Biochemistry of the platelet release reaction, in "Biochemistry and Pharmacology of Platelets", Ciba Founda-

tion Symposium, Elsevier North Holland Biomedical Press, Amsterdam, p. 175 (1975).

17. H. Holmsen, The platelet, its membrane physiology and biochemistry, in "Clinics in Hematology", W.B. Sanders Co., London, 2:235 (1972).

18. M.E. Bentfield and D.F. Bainton, Cytochemical localization of lysosomal enzymes in rat megakaryocytes and platelets, J. Clin. Invest., 56:1635 (1975).

19. R.V. Buker, H. Blaschko and G.V.R. Born, The isolation from blood platelets of particles containing 5 hydroxy tryptamine and adenosine diphosphate, J. Physiol., 149:55 (1959).

20. R.B. Davis and G.C. White, Localization of 5-hydroxytryptamine in blood platelets, Brit. J. Haematol., 15:93 (1968).

21. M.J. Broekman, R.I. Handin and P. Cohen, Distribution of fibrinogen and platelet factor 4 an XIII in subcellular fractions of human platelets, Brit. J. Haematol., 31:51 (1975).

22. K.I. Kaplan, J. Kaplan, J. Broekman, A. Chernoff, G.R. Lezzink, and H. Drillings, Platelet alpha granule proteins: studies on release and subcellular localization, Blood, 53:604 (1979).

23. B. Rucinski, S. Newiarowski, P. James, D.A. Walz, and A.Z. Budzinski, Antiheparin proteins secreted by human platelets. Purification, characterization and radioimmunoassay, Blood, 53:47 (1979).

24. J.F. Mustard, Z. Movat, D.R.L. Mac Morine and A. Senyi, Release of permeability factors from blood platelets, Proc. Soc. Exptl. Biol. Med., 119:988 (1965).

25. R. Ross, J. Glomset, B. Kariya and L. Harker, A platelet dependent serum factor that stimulates the proliferation of arterial smooth muscle in vitro, Proc. Natl. Acad. Sci. USA, 71:1207 (1974).

26. A. Siegel and E.F. Luscher, Non identity of the granules of human blood platelets with typical lysosomes, Nature, 215:745 (1967).

27. M.B. Donati and G. de Gaetano, Interaction of fibrin with blood platelets and other cells, in "Comparison of Life Sciences", vol. 2, B.A. Warner, ed., Van Nostrand Reinhold Co, New York, N.Y., USA, (1981).

28. P.N. Walsh, The effects of collagen and kaolin on the intrinsic coagulant activities of platelets, Brit. J. Haema-

       <u>tol.</u>, 22:393 (1972).

29. T.H. Spaet and J. Cintron, Studies on platelet factors bio-
    availability, <u>Brit. J. Haematol.</u>, 11:269 (1965).

30. P.N. Walsh, R.E. Goldberg, R.L. Tax, and L.E. Margadal, Plate-
    let coagulant activities and retinal vein thrombosis,
    <u>Thromb. and Haemost.</u>, 38:399 (1977).

31. N. Semeraro and J. Vermylen, Evidence that washed human
    platelets possess factor x activator activity, <u>Brit. J.</u>
    <u>Haematol.</u>, 11:269 (1965).

32. E. Tremoli, M.B. Donati and G. de Gaetano, Washed guinea pig
    and rat platelets possess factor x activating activity,
    <u>Brit. J. Haematol.</u>, 9:155 (1977).

33. E. Tremoli, M. Colucci, M.B. Donati and N. Semeraro, Early
    increase of a new platelet coagulant activity in rats fed
    a thrombogenic diet, <u>Atherosclerosis</u>, 33:239 (1979).

34. J. Kloeze, Influence of prostaglandins on platelet adhesive-
    ness and platelet aggregation, <u>in</u> "Prostaglandins, Pro-
    ceedings of the 2nd Noble Symposium", S. Bergstrom and B.
    Samuelsson eds., Sune, New York, Interscience (1966).

35. S. Bergstrom, H. Danielsson, and B. Samuelsson, The enzymatic
    formation of prostaglandin E_2 from arachidonic acid, <u>Bio-</u>
    <u>chim. Biophys. Acta</u>, 90:207 (1964).

36. R.J. Gryglewski, J.A. Salmon, F.B. Ubatuba, B.C. Weatherly,
    S. Moncada, and J.R. Vane, Effects of all cis-5,8,11,14,
    17-eicosapentaenoic acid and PGH_3 on platelet aggrega-
    tion, <u>Prostaglandins</u>, 18:453 (1979).

37. J.B. Smith and A.L. Willis, Formation and release of prosta-
    glandins in response to thrombin, <u>Br. J. Pharmacol.</u>,
    40:545 (1970).

38. H. Shio and P.W. Ramwell, Effect of prostaglandin E_2 and
    aspirin on the secondary aggregation of human platelets,
    <u>Nature</u>, 236:45 (1972).

39. M.J. Silver, J.B. Smith, C. Ingerman, and J.J. Kosics,
    Arachidonic acid induced human platelet aggregation and
    prostaglandin formation, <u>Prostaglandins</u>, 4:863 (1973).

40. M. Hamberg and B. Samuelsson, Detection and isolation of an
    endoperoxide intermediate in prostaglandin biosynthesis,
    <u>Proc. Natl. Acad. Sci. USA</u>, 70:899 (1973).

41. M. Hamberg, J. Svensson, and B. Samuelsson, Thromboxanes a
    new group of biologically active compounds derived from
    prostaglandin endoperoxides, <u>Proc. Natl. Acad. Sci. USA</u>,
    72:2994 (1975).

42. G. Bereziat, Are phospholipases involved in platelet activa-
    tion?, Agents and Action, 9:390 (1979).
43. S. Hammarstrom, Selective inhibition of platelet n-8 lipooxy-
    genase by 5,8,11-eicosatryenoic acid, Biochim. Biophys.
    Acta, 487:517 (1977).
44. B. Samuelsson, P. Borgeat, S. Hammarstrom, and R.C. Murphy,
    Leukotrienes: a new group of biologically active com-
    pounds, in "Adv. in Prostaglandin and Thromboxane Re-
    search", vol. 6, B. Samuelsson, P.W. Ramwell, R. Paoletti,
    eds., Raven Press, New York, N.Y., p. 1 (1980).
45. A.W. Ford-Hutchinson, M.A. Bray, H.Y. Doig, M.E. Shepley, and
    M.J.H. Smith, Leukotriene B, a potent chemokinetic and
    aggregating substance released from polymorphonuclear
    leukocytes, Nature, 286:264 (1980).
46. H.J. Weiss, L.M. Aledort, and A.S. Koch, The effect of sali-
    cylates on the hemostatic properties of platelets in man,
    J. Clin. Invest., 47:2169 (1968).
47. J.B. Smith, M.J. Silver, C.M. Ingerman, and J.J. Kosics, Pros-
    taglandin D_2 inhibits the aggregation of human platelets,
    Thromb. Res., 5:291 (1974).
48. J.B. Smith, A.W. Sedar, C.M. Ingerman, and M.J. Silver, Pros-
    taglandin endoperoxides: Platelet shape change, aggrega-
    tion and the release reaction, in "Platelets and Thrombo-
    sis", D.C.B. Mills and F.I. Pareti eds., p. 83, Academic
    Press, New York, N.Y. (1977).
49. J. Svensson, M. Hamberg, and B. Samuelsson, On the formation
    and effects of thromboxane A_2 on human platelets, Acta
    Physiol. Scand., 98:285 (1976).
50. S. Moncada, R.J. Gryglewski, S. Bunting, and J.R. Vane, An
    enzyme isolated from arteries transforms prostaglandin
    endoperoxides to an unstable substance that inhibits
    platelet, Nature, 263:663 (1976).
51. S. Moncada, A.G. Herman, E.A. Higgs, and J.R. Vane, Differen-
    tial formation of prostacyclin (PGX or PGI_2) by layers of
    the arterial wall. An explanation for the antithrombotic
    properties of vascular endothelium, Thromb. Res., 3:323
    (1977).
52. S. Moncada, J.R. Vane, and B.J.R. Whittle, Relative potency
    of Prostacyclin, Prostaglandin E_1 and D_2 as inhibitors of
    platelet aggregation in several species, J. Physiol.,
    273:2P (1977).
53. R.J. Gryglewski, R. Korbut, and A. Ocetkiewicz, Reversal of
    platelet aggregation by prostacyclin, Pharm. Res. Comm.,

10:185 (1978).

54. N.G. Ardlie, G. Glew, B.G. Schultz, and C.J. Schwartz, Inhibition and reversal of platelet aggregation by methylxanthines, Thromb. Diath. Haemorrh., 18:670 (1967).

55. R.J. Haslam, Interaction of the pharmacological receptors of blood platelets with adenylate cyclase, Ser. Haematol., 6:333 (1973).

56. E.W. Salzman, Cyclic AMP and platelet function, N. Engl. J. Med., 286:358 (1972).

57. E.W. Salzman, Prostaglandins, Cyclic AMP and platelet function, Thromb. Diath. Haemorrh., 60 (suppl.):311 (1974).

58. E.W. Salzman, J. Lindon, D. Brier, and E.W. Merrill, Surface-induced platelet adhesion, aggregation and release, Ann. N.Y. Acad. Sci. 283:114 (1977).

59. D.C.B. Mills and J.B. Smith, The influence on platelet aggregation of drugs that affect the accumulation of adenosine 3'5'-cyclic monophosphate in platelets, Biochem. J., 121: 185 (1971).

60. N.D. Goldberg, R.F. O'Dea, and M.K. Haddox, Cyclic GMP, Adv. Cyclic Nucleotide Res., 3:155 (1973).

61. L.C. Best, T.J. Martin, R.G.G. Russell, and F.E. Preston, Prostacyclin increases cyclic AMP levels and adenylate cyclase activity in platelets, Nature, 267:850 (1977).

62. D.C.B. Mills and D.E. MacFarlane, Stimulation of human platelet adenylate cyclase by prostaglandin D_2, Thromb. Res., 5:401 (1974).

63. R.R. Gorman, S. Bunting, and O.V. Muller, Modulation of human platelet adenylate cyclase by prostacyclin (Pgx), Prostaglandins, 13:377 (1977).

64. G.C. LeBreton, D.L. Venton, S.E. Enke, and P.V. Halushka, 13-Azaprostanoic acid: a specific antagonist of human blood platelet thromboxane/endoperoxide receptor, Proc. Natl. Acad. Sci. USA, 76:4097 (1979).

65. D.L. Venton, S.E. Enke, and G.C. Le Breton, Azaprostanoic derivatives. Inhibitors of arachidonic and induced platelet aggregation, J. Med. Chem., 22:824 (1979).

66. A.M. Siegl, J.B. Smith, M.J. Silver, K.C. Nicolaou, G. Gasic, and W.E. Burnette, Binding of prostacyclin by platelets, Federation Proceedings, 37:260 (1978).

67. G. Di Minno, M.J. Silver, and G. de Gaetano, Prostaglandins as inhibitors of human platelet aggregation, Brit. J. Hematol., 43:637 (1979).

68. J. Mehta and P. Mehta, Enhanced platelet prostaglandin genera-
    tion and abnormal platelet sensitivity to prostacyclin
    and thromboxane A$_2$ in angina pectoris, Blood, 54 (sup-
    pl.):2510 (1979).
69. J.F. Mustard and M.A. Packham, The role of blood platelets in
    atherosclerosis and the complications of atherosclerosis,
    Thromb. Diath. Haemorrh., 33:444 (1975).
70. J.A. Colwell, P.V. Halushka, K.E. Sarji, and J. Sagel, Plate-
    let function and diabetes mellitus, Med. Clin. North Am.,
    62:753 (1978).
71. P.V. Halushka, D. Lurie, and J.A. Colwell, Increased synthe-
    sis of prostaglandin E-like material by platelets from pa-
    tients with diabetes mellitus, N. Engl. J. Med., 297:1306
    (1977).
72. M. Lagarde and M. Dechavanne, Increase of platelet prostaglan-
    din cyclic endoperoxides in thrombosis, Lancet, 1:88
    (1977).
73. M. Lagarde and M. Dechavanne, Collagen stimulates more plate-
    let prostaglandin endoperoxides in thrombosis, Biomedi-
    cine, 27:119 (1977).
74. WHO Memorandum, Classification of hyperlipidemias and hyper-
    lipoproteinemias, Circulation, 45:501 (1975).
75. D.S. Fredrickson, R.I. Levis, and R.S. Lees, Fat trasnport in
    lipoproteins. An integrated approach to mechanisms and
    disorders, N. Engl. J. Med., 276:32,94,148,215,273 (1967).
76. K. Westlund and R. Nicolaysen, Serum cholesterol and risk of
    mortality and morbidity, Scand. J. Clin. Lab. Invest., 87
    (suppl.) (1966).
77. A. Norddy and J.M. Rodset, Platelet function and platelet
    phospholipids in patients with hyperlipoproteinemia, Acta
    Med. Scand., 189:385 (1971).
78. A. Carvalho, R.W. Colman, and R.S. Lees, Platelet function in
    hyperlipoproteinemia, N. Engl. J. Med., 290:434 (1974).
79. E. Tremoli, P. Maderna, M. Sirtori, and C.R. Sirtori, Plate-
    let aggregation and malondialdehyde formation in type IIA
    hypercholesterolemic patients, Haemostasis, 8:47 (1979).
80. E. Tremoli, G.C. Folco, E. Agradi, and C. Galli, Platelet
    thromboxanes and serum cholesterol, Lancet, 1:107 (1979).
81. A. Zmuda, A. Dembinska-Kiec, A. Chytocwski, and R.J. Gryglew-
    ski, Experimental atherosclerosis in rabbits-platelet ag-
    gregation, Thromboxane A$_2$ generation and anti-aggregating
    potency of prostacyclin, Prostaglandins, 16:1035 (1977).

82. S.J. Shattil, R. Anaya-Galindo, J. Bennett, R.W. Colman, and R.A. Cooper, Platelet hypersensitivity induced by cholesterol incorporation, J. Clin. Invest., 55:636 (1975).

83. M.J. Stuart, J.M. Gerrard, and J.G. White, Effect of cholesterol on production of thromboxane B_2 by platelets in vitro, N. Engl. J. Med., 302:6 (1980).

84. J.W. Burch, N. Stanford and P.W. Majerus, Inhibition of platelet prostaglandin synthetase by oral aspirin, J. Clin. Invest., 61:314 (1978).

85. M. Buchanan and J. Hizsch, Comparison of in vivo and in vitro effects of platelet function suppressing drugs, in "Proceedings of the 5th Congress of the Intern. Society of Thrombosis and Haemostasis", Paris, abstract 182 (1979).

86. S. Moncada, P. Korbut, S. Bunting, and J.R. Vane, Prostacyclin is a circulating hormone, Nature, 273:767 (1978).

87. S. Moncada and J.R. Vane, Prostacyclin in the cardiovascular system, in "Adv. in Prostaglandin and Thromboxane Research", vol. 6, B. Samuelsson, P.W. Ramwell and R. Paoletti, eds., Raven Press, New York, N.Y. (1980).

88. N.L. Baenziger, M.J. Dillender, and P.W. Majerus, Cultured human skin fibroblasts and arterial cells produce a labile platelet-inhibitory prostaglandin, Biochem. Biophys. Res. Comm., 78:294 (1977).

89. J.C. Kelton, J. Hirsch, C.J. Carter, and M.R. Buchanan, Thrombogenic effect of high-dose aspirin in rabbits, J. Clin. Invest., 62:892 (1978).

90. C. Patrono, G. Ciabattoni, E. Pinca, F. Pugliese, G. Castrucci, A. De Salvo, M.A. Satta, and B.A. Peskar, Low dose Aspirin and inhibition of thromboxane B_2 production in healthy subjects, Throm. Res., 17:317 (1980).

91. S. Villa, M. Livio, and G. de Gaetano, The inhibitory effect of aspirin on platelet and vascular prostaglandins in rats cannot be completely dissociated, Brit. J. Haematol. 42:425 (1979).

92. G. Masotti, G. Galanti, S. Poggesi, R. Abbate, and G.G. Neri Serneri, Differential inhibition of prostacyclin production and platelet aggregation by aspirin, Lancet, ii:1213 (1979).

93. E. Tremoli, P. Maderna, M. Sirtori, S. Colli, G. Corvi, and C.R. Sirtori, Indobufen therapy in type IIa hypercholesterolemic subjects: effects on platelet function and malondialdehyde production, Haemostasis, (in press).

94. P. Mehta and J. Mehta, Platelet function studies in coronary heart disease. VIII. Decreased platelet sensitivity to prostacyclin in patients with myocardial ischaemia, Thromb. Res., 18:273 (1980).

95. J. Merino, M. Livio, G. Rajtar, and G. de Gaetano, Salicylate reverses in vitro aspirin inhibition of rat platelet and vascular prostaglandin generation, Biochem. Pharmacol., 29:1093 (1980).

96. S. Riegelman, The kinetic disposition of aspirin in human, in "Aspirin, Platelets and Stroke", Background for a Clinical Trial, W.S. Fields and W.K. Hass, eds., p. 105, Wanen H. Green, St. Louis (1971).

97. M.B. Zucker and J. Peterson, Effect of acetylsalicylic acid, other nonsteroidal antiinflammatory agents and dipyridamoleon human blood platelets, J. Lab. Clin. Med., 76:66 (1970).

98. D.C.B. Mills and J.B. Smith, The influence on platelet aggregation of drugs that affect the accumulation of adenosine 3',5' Cyclic monophosphate in platelets, Biochem. J., 121:185 (1971).

99. S. Moncada and R. Korbut, Dipyridamole and other phosphodiesterase inhibits act as antithrombotic agents by potentiating endogenous prostacyclin, Lancet, i:1286 (1978).

100. L.A. Harker and S.J. Slichter, Studies of platelet and fibrinogen kinetics in patients with prosthetic heart valves, N. Engl. J. Med., 283:1302 (1970).

101. L.A. Harker, R. Ross, S.J. Slichter, and C.R. Scott, Homocistine induced aterosclerosis. The rate of endothelial cell injury and its response on his genesis, J. Clin. Invest., 58:731 (1976).

102. B.J.R. Whittle, S. Moncada, F. Whiting, and J.R. Vane, Carbacyclin, a potent stable prostacyclin analogue for the inhibition of platelet aggregation, Prostaglandins, 19:605 (1980).

103. L.D. Tobias and J.G. Hamilton, Inhibition of Arachidonate metabolism by selected compounds in vitro with particular emphasis on the thromboxane A_2 synthase pathway, in "Adv. in Prostaglandin and Thromboxane Research", vol. 6, B. Samuelsson, P.W. Ramwell and R. Paoletti, eds., Raven Press, New York, N.Y. (1980).

104. J. Vermylen, D.A.F. Chamone, and M. Verstraete, Stimulation of prostacyclin release from vessel wall by BAY g 6575, an antithrombotic compound, Lancet, ii:518 (1979).

105. S. Ashida and Y. Abiko, Effect of ticlopidine and acetyl salicylic acid on generation of prostaglandin I_2-like substance in rat arterial tissue, Thromb. Res., 13:901 (1978).

106. W.S. Fields, N.A. Lemak, R.F. Frankowski, and R.L. Hardy, Controlled trial of aspirin in cerebral ischemia, Stroke, 8:301 (1977).

107. The Canadian Cooperative Study Group, A randomized trial of aspirin and sulfinpyrazone in threatened stroke, N. Engl. J. Med., 299:53 (1978).

108. Aspirin Myocardial Infarction Study Research Group, A randomized, controlled trial of aspirin in persons recovered from myocardial infarction, JAMA, 243:661 (1980).

109. P.C. Elwood, A.L. Cochrane, M.L. Burr, P.M. Sweetman, G. Williams, S.J. Hughes, and R. Renton, A randomized controlled trial of acetylsalicylic acid in the secondary prevention of mortality from myocardial infarction, Br. Med. J., 1:436 (1974).

110. The Coronary Drug Project Research Group, Aspirin in coronary heart disease, J. Chronic Dis., 29:625 (1976).

111. The Anturane Reinfarction Trial Research Group, Sulfinpyrazone in the prevention of cardiac death after myocardial infarction, N. Engl. J. Med., 296:289 (1978).

112. The Persantin-Aspirin Reinfarction Study Research Group, Persantin and aspirin in coronary heart disease, Circulation, 62:449 (1980).

113. J. Stamler, Diet-related risk factors for human atherosclerosis: hyperlipidemia, hypertension, hyperglycemia - current status, Adv. Exp. Med. Biol., 60:125 (1975).

114. R.W. Wissler and D. Vesselinovitch, The effects of feeding various dietary fats on the development and regression of hypercholesterolemia and atherosclerosis, Adv. Exp. Med. Biol., 60:65 (1975).

115. S. Renaud, C. Allard, and J.G. Latour, Influence of the type of dietary saturated fatty acid on lipemia, coagulation and the production of thrombosis in the rat, J. Nutr., 90:433 (1966).

116. G. Hornstra, Specific effects of types of dietary fat on arterial thrombosis, in "The role of fats in human nutrition", A.J. Vergroesen ed., Academic Press, New York, N.Y., p. 303 (1975).

117. L. McGregor and S. Renaud, Effect of dietary linoleic acid deficiency on platelet aggregation and phospholipid fatty

acids of rats, Thromb. Res., 12:921 (1978).

118. C. Galli, C. Spagnuolo, E. Bosisio, L. Tosi, G.C. Folco, and G. Galli, Dietary essential fatty acids brain polyunsaturated fatty acids and prostaglandins in the central nervous system, in "Adv. Prostaglandins and Thromboxane Research", vol. IV, F. Coceani and P.M. Olley eds., Raven Press, New York, N.Y., p. 181 (1978).

119. E. Agradi, E. Tremoli, C. Colombo, and C. Galli, Influence of short term dietary supplementation of different lipids on aggregation and arachidonic acid metabolism in rabbit platelets, Prostaglandins, 16:973 (1978).

120. J. Dyerberg, H.O. Bang, E. Stoffersen, S. Moncada, and J.R. Vane, Eicosapentaenoic acid and prevention of thrombosis and atherosclerosis, Lancet, 2:117 (1978).

121. P. Needleman, H. Sprecher, M.O. Whitaker, and A. Wyche, Mechanism underlying the inhibition of platelet aggregation by eicosapentaenoic acid and its metabolites, in "Adv. in Prostaglandin and Thromboxane Research", vol. 6, B. Samuelsson, P.W. Ramwell and R. Paoletti, eds., Raven Press, New York, N.Y., p. 61 (1980).

122. A. Raz, M.S. Minkes and P. Needleman, Endoperoxides and thromboxanes. Structural determinants for platelet aggregation and vasoconstriction, Biochim. Biophys. Acta, 488: 305 (1977).

123. M.O. Whitaker, P. Needleman, A. Wyche, F.A. Fitzpatrick, and H. Sprecher, PGD_3 is the mediator of the antiaggregatory effects of the trienoic Endoperoxide PGH_3, in "Adv. in Prostaglandin and Thromboxane Research", vol. 6, B. Samuelsson, P.W. Ramwell and R. Paoletti, eds., Raven Press, New York, N.Y., p. 301 (1980).

124. P. Needleman, M.O. Whitaker, A. Wyche, K. Watten, H. Sprecher, and A. Raz, Manipulation of platelet aggregation by prostaglandins and their fatty acid precursors: pharmacological basis for a therapeutic approach, Prostaglandins, 19: 165 (1980).

125. S. Hammarstrom, Leukotriene C_5: a slow reacting substance derived from EPA, J. Biol. Chem., 255:7093 (1980).

TOBACCO SMOKING AND ATHEROSCLEROSIS

Constantin J. Miras

Department of Medicine, University of Athens, Goudi -

Athens, Greece

Extensive epidemiological data associate tobacco smoking with
an increased risk for atherosclerosis (1-5). Beyond any doubt morbidity and mortality, especially from coronary heart disease, are
increased in smokers (6-7), more often as precipitating factor
of the terminal occlusive or thrombotic episode on a preexisting
atherosclerotic growth. Although this contribution is defined (8-9),
a full explanation is lacking.

Less documented, tobacco smoking has been reported also as a
risk factor for cerebral thrombosis (10).

Table 1. Estimated years of life expectancy (LE) for males at various
ages by amount of cigarettes smoked, U.S. Veterans Study(11)

Cigarettes smoked per day	Age							
	35		40		50		60	
	LE	YL[o]	LE	YL	LE	YL	LE	YL
Nonsmokers	43.5	0	38.7	0	29.4	0	20.8	0
1-10	41.0	2.5	36.3	2.4	27.5	1.9	19.0	1.8
10-20	38.7	4.8	34.1	4.6	25.2	4.2	17.2	3.6
21-39	36.7	6.8	32.0	6.7	23.4	6.0	15.8	5.0
40[+]	34.8	8.7	29.9	8.8	21.6	7.8	14.4	6.4

YL[o] = Years lost

Women smokers have the same risk as man (12). Heavy smoking
has also been found to produce changes in the umbilical cord (13),
the placenta and the foetus (14-15). The mechanism by which smoking
is involved in atherosclerosis has not been elucidated.

Heavy smoking is usually an expression of anxiety and psycho-
social stress in individuals with a particular personality. The
possibility that this type of personality may be the linked association that has been measured in epidemiological studies has to be
considered as well.

However, in Greece many tobacco cultivators* are heavy smokers
traditionally and not as a symptom of a particular personality. They
have also higher incidence of cardiovascular diseases than the non-
smokers in the same area (16).

Although it has been reported that tobacco smoke contains about
4000 substances, only few of the tobacco smoke constituents have
been tested.

Some of the substances included in Table II have pharmacological,
mitogenic, allergenic and toxic effects. Polycyclic hydrocarbon has
been reported to enhance spontaneous atherogenesis in chicken (18).
An increased activity of aryl hydrocarbon hydroxylase has been found
in the umbilical cord of heavy smokers (13) throughout the gestation
period.

The problem is complicated, since the quality and quantity of
different substances in tobacco smoke varies significantly. The
variation depends not only on the quality of tobacco, but also on
the dryness of smoked tobacco and the way of smoking, that affects
the pyrolysis of the different compounds. On the other hand, the
amount inhaled depends also on the depth of inhalation, that is
difficult to be evaluated (19).

An increase of erythrocytes, hematocrit, and hemoglobin has
been reported in chronic smokers. Also some slight changes in the
biochemical profile (blood sugar, fatty acids, etc.). It seems
however, that the above changes are not important for the pathogen-

---

* Tobacco cultivators in Greece smoke their own product tax exempt-
ed. So, smoking is rather a traditional habit to "enjoy" something
(for nothing) that others have to pay. To some extent the same has
been observed in tobacco industry workers.

Table II. Some of the most important constituents of tobacco smoke (17)

Gas Phase	Particulate Phase
Carbon monoxide	Nicotine
Carbon dioxide	Tar
Nitrogen oxides	Non-volatile N-nitrosamines
Ammonia	Aromatic amines
Acrolein acid-Methacrolein	Isoprenoids
Volatile N-nitrosamines	Pyrenes
Hydrogen cyanide	Benzopyrenes
Volatile sulfur compounds	Chrysenes
Nitriles	Anthracenes
Other volatile nitrogen	
containing compounds	Fluoranthenes
Volatile hydrocarbons	Carcinogenic aza-arenes
Alcohols	a. acridines
Aldehydes (acetaldehyde)	b. carbazoles
Ketones (acetone)	c. quinolines
Acetonitrile	d. phenanthridines
Benzene	e. other minor components
	Phenols
	Cresols
	Naphthols
	Alkanes and alkenes
	Naphthalenes
	Carboxylic acids
	Metallic ions

esis of atherosclerosis, except this of the increased Low Density
Lipoprotein Cholesterol (LDL-C), at the expenses of High Density
Lipoprotein Cholesterol (HDL-C). This disturbance in lipoprotein
pattern is generally accepted as a risk factor for atherosclerosis.

Atherosclerosis is a multifactorial disease that produces, with
some selectivity, a large number of arterial wall lesions,affecting
different parts of the vascular system. The possibility that tobacco
smoke constituents may act independently from other risk factors
on the arterial wall cannot be excluded.

Some evidence has been recently presented concerning the decreased integrity of arterial endothelium of uterine arteries in women
smokers (20). This possibility needs further investigations.

Tobacco smoke constituents may influence the development of
atherogenesis, potentiating other factors' effects, interfering
with repair mechanism and metabolism of the arterial wall.

One of the most extensively studied constituents of tobacco
smoke is CO. High levels of Hb-CO in blood are implicated with pathogenesis of atherosclerosis (21-23). But levels usually found in smokers
are not considered today as a very important factor, except if they
are connected with hypercholesterolemia (24), because non--smokers
working in polluted atmosphere have some times quite high Hb-CO
levels in blood without evidence of atherosclerosis (Table III).

Most of the studies with CO have been done on animal models
and the extrapolation to humans is not easy.

Table III. Carboxyhemoglobin levels in smokers and non-smokers
           (Mean range of Hb-CO%)

		Non-smokers	Smokers
Office workers	(25)	1.1	3.6
London office workers	(26)	1.12	5.5
		(0.1-2.7)	(2.2-13.0)
USA blood donors	(27)	1.39	5.57
(29.000)		(0.4-6.9)	(0.8-11.9)
Munich population	(28)	2.36	7.38
Munich population	(29)	2.98	4.87
		(1.5-4.9)	(2.8-9.6)
Rural Bavarians	(28)	1.03	6.06

A direct effect of tobacco smoking attributed to the CO has been
reported on hepatic proteinsynthesis (30). To what extent this effect
can be connected to changes in the lipoprotein pattern of heavy
smokers is not known. Carboxyl-hemoglobin decreases also chylomicron
remnants catabolism by the liver (31) and the delayed remnant clearance may potentiate atherogenesis, as in type III hyperlipoproteinemias. However chronic exposure to cigarette smoke constituents may
cause induction (increased biosynthesis) of drug metabolizing enzymes

of the liver. This observation is in connection with increased
activity of liver microsomal mixed function monooxygenases. To
what extent this activity can influence the metabolism of hormones
by the liver or the metabolism of different substances of tobacco
smoke has not been proved yet.

Recently Aronow (33) found that no-nicotine cigarettes produce
earlier exercise-induced angina on humans than the CO contained in
the smoked cigarettes. It is obvious that substances in tobacco
smoke other than CO and nicotine contribute to the observed effect.

In any case, Hb-CO level in blood is a valuable marker of
tobacco smoking and a good index to associate smoking effects (34).

Nicotine is the next well studied substance of cigarette smoke
that produces the well known hemodynamic and metabolic dose-related
effects. The release of catecholamines is responsible for the observed phenomena. Spohr et al. (35) have recently studied the tobacco
smoke sympathetic activity, by measuring the Dopamine β-Hydroxylase
(DBH). They reported an increased activity of DBH as well as other
biochemical parameters connected with increased sympathetic activity.
The increase of DBH is very interesting and although connected with
the nicotine effect, it is probably influenced directly or indirectly
by other substances.

Detailed studies determining the consumption of O_2 by the myocardium, prove the presence of inotropic and chronotropic substances
other than nicotine (36).

Table IV. Blood levels of nicotine in cigarette smokers

	Nicotine content (ng/ml)	Methods
Isaac and Rand (37)	25	GLC venous (plasma)
Armitage et al.(38)	30–40	^{14}C-labeled nicotine (arterial whole blood)
Hill (39)	38	Radioimmunoassay (serum)

The positive inotropic and chronotropic effect of nicotine on
myocardium,due to the release of catecholamines and corticosteroids,
aggravates the coronary heart disease and influences the development
of the clinical picture (40-45).

In addition to the other effects release of catecholamines may
contribute to a thrombous formation on atherosclerotic lesions
through an effect on platelets. A number of observations confirm
the effect of tobacco smoke on platelets functions, in particular
platelet aggregation (46-50).

Such effect may contribute to thrombous formation in smokers. However smoking enhances coronary, cerebral and limb arteries thrombosis but not venous thrombosis (51). It is obvious that the interaction of the sensitized platelet with damged vascular bed is of critical importance. Activation of factor XII by a tobacco constituent has also been reported (52).

The presence of allergenic substances in tobacco smoke is well recognized. An immunological reaction may injure the arterial wall and contribute to pathogenesis of atherosclerosis (53-55). Becker has isolated a glycoprotein (M.W. 18000-26000) from tobacco leaves that produces cutaneous hypersensitivity in a good percentage of volunteers (56).

An induction of IgE antibodies to this antigen has been found (57). Changes in endothelium permeability can follow IgE reactions and can explain arterial damage of hypercholesterolemia. Such an observation has been reported on swine by Chopath (58).

Undoubtely much more data are needed in order to elucidate the actual role of immunological reactions in the development of spontaneous atherosclerosis in smokers.

A number of epidemiological studies (59-61) have shown an alteration of lipoprotein pattern in smokers. It is now generally accepted that smoking increases the LDL (atherogenic) and decreases HDL (antiatherogenic)(62). Total serum Ch may only slightly be increased because of the opposing effect on the two main Ch carrying lipoproteins and the greater influence of smoking on LDL. In addition to the above observations Apo-A$_I$ levels has been found to be lower in HDL of smokers in comparison with those of non-smokers (63).

Since the Apo-A$_I$ is the main Apo protein of the HDL family, the above finding is consistent with the previous mentioned effect of smoking on HDL. The mechanism by which tobacco smoke constituents may influence lipoprotein level is only a speculative one.

Conclusions and Summary

Tobacco smoking definitely contributes to the atherosclerotic disease. The mechanism is not clear although many substances of the smoke are candidates as risk factors. It is obvious that more than one substance from tobacco smoke contribute by different ways in all stages of the disease. Direct effect on the vascular bed is probable for allergenic or mitogenic agents. However the development of atherosclerotic lesion through the change in permeability and lipoproteins pattern is suspected.

The contribution of tobacco smoking in the terminal occlusive or thrombotic episode is very well documented. Undoubtely any agent that can interfere with the arterial metabolism or increases the workload of the heart may contribute to the clinical picture of atherosclerotic disease without any direct effect on atherogenesis.

Almost 4000 substances have been reported to be in the tobacco smoke, but only few of them have been tested for atherogenic activity. The same is true for their active metabolites. The effect of CO and nicotine don't explain the association between atherogenesis and smoking.

Very few work has been done in this field with tobacco smoke antigens and the immune complex reactions in the arterial wall.

The same is true for the mitogenic substances.

It is obvious that the research must be conducted towards this direction.

References

1. W.B. Kannel, P. Sorlie, F. Bran, W.P. Castelli, P.M. McNamara, and G.J. Gherardi, Epidemiology of Coronary Atherosclerosis: Postmortem VS Clinical Risk Factor Correlations,The Framingham Study, in "Atherosclerosis V. Proceedings of the Fifth International Symposium", A.M. Gotto,Jr., L.C. Smith and B. Allen, eds., Springer-Verlag, New York (1980)

2. G.N. Stemmermann, G.G. Rhoads, and T. Hayashi, Atherosclerosis and Its Risk Factors Among Hawaii Japanese, in "Atherosclerosis V. Proceedings of the Fifth International Symposium" ,A.M. Gotto,Jr.,L.C. Smith and B. Allen,eds.,Springer-Verlag, New York (1980)

3. N.H. Sternby, Atherosclerosis, Smoking and Other Risk Factors, in "Atherosclerosis V. Proceedings of the Fifth International Symposium", A.M. Gotto,Jr.,L.C. Smith and B. Allen,eds., Springer-Verlag, New York (1980)

4. J.P. Strong and P. Omae, Epidemiology of Atherosclerosis and Geographic Differences in Risk Factors", in "Atherosclerosis IV. Proceedings of the Fourth International Symposium, G. Schettler, Y. Goto, Y. Hata, G. Klose,eds.,Springer-Verlag, Berlin (1977)

5. J.P. Strong and M.L. Richards, Cigarette smoking and atherosclerosis in autopsied men, Atherosclerosis 23(3):451 (1976)

6. R.L. Naeye and L.D. Truong, Effects of cigarette smoking on intramyocardial arteries and arterioles in man, Am.J. Clin.Path. 68(4):493 (1977)

7. O. Auerbach, H.W. Carter, L. Garfinkel and E.C. Hammond, Cigarette smoking and coronary artery disease: A macroscopic and microscopic study, Chest 70(6):697 (1976)

8. J.T. Salonen, Stopping smoking and long-term mortality after acute myocardial infarction, Br.Heart J. 43:463 (1980)

9. W.J. McKenna, C.Y.C. Chew and C.M. Oakley, Myocardial infarction with normal coronary angiogram. Possible mechanism of smoking risk in coronary artery disease, Br. Heart J. 43:493 (1980)

10. W.B. Kannel, Epidemiological studies on smoking in cerebral and peripheral vascular disease, in "Smoking and Health.I. Modifying the Risk for the Smoker", E. Wynder, D. Hoffmann and G.B. Gori,eds., DHEW Publication No. (NIH) 76-1221, (1975)

11. E. Rogot, Smoking and General Mortality Among U.S. Veterans, 1954-1969, U.S.A. Department of Health, Education and Welfare, Public Health Service, National Institutes of Health, National Heart and Lung Institute, Epidemiology Branch, DHWE Publication No. (NIH) 74-544 (1974)

12. R. Doll, R. Gray, B. Hafner and R. Peto, Mortality in relation to smoking:22 years observations on female British doctors, Brit.Med.J. 281:967 (1980)

13. C.J. Miras, J. Lekakis, A. Bouloukos and A. Kalofoutis, Surface cellular fibronectin, aryl hydrocarbon hydroxylase in umbilical cord of heavy smokers at delivery, Preliminary report submitted for publication

14. R.L. Naeye, Effects of maternal cigarette smoking on foetus and placenta, Brit.J.Obstetr.Gynaecol. 85(10):732 (1978)

15. O. Pelkonen, N.T. Karki, M. Koivisto, R. Tuimala and A. Kauppila, Maternal cigarette smoking, placental aryl hydrocarbon hydroxylase and neonatal size, Toxicol.Letters ,NLD,3,No.6,331 (1979)

16. C.J. Miras and A. Kalofoutis, Tobacco smoking as a risk factor in olive oil consume areas (Lesvos,Laconia), Proceedings of the Pankretion Medical Congress, Rethymnon-Kreta (1980)

17. A report of Surgeon General. Reprint. Constituenst of Tobacco Smoke, smoking and health, D.C. Governement Printing Office, pp. 33-70, Washington (1979)

18. R.E. Albert, M. Vanderlaan, F.J. Burns and M. Nishizumi, Effect of carcinogens on chicken atherosclerosis, Cancer Res. 37(7):2232 (1977)

19. L.C. Miller, A.F. Schilling, D.L. Logan, et al., Potential hazards of rapid smoking as a technic for the modification of smoking behaviour, New Eng.J.Med. 297:590 (1977)

20. A. Bylock, G. Bondjiers, I. Jansson and H.A. Hansson, Surface
    ultrastructure of human arteries with special reference to the
    effects of smoking, Acta Path.Microbiol.Scand.,Section A.,
    87:201 (1979)

21. P. Astrup, Some physiological and pathological effects of moderate
    carbon monoxide exposure, Brit.Med.J. 2:447 (1972)

22. N. Wald, S. Howard, P.G. Smith and K. Kjelsen, Association between
    atherosclerotic diseases and carboxyhemoglobin levels in tobacco
    smokers, Brit.Med.J. 31p:761 (1973)

23. M. Heliövaara, M.J. Karvonen, R. Vilhurien and S. Punsar, Smoking,
    carbon monoxide, and atherosclerotic diseases, Brit.Med.J. 1:268
    (1978)

24. D.M. Turner, Carbon monoxide tobacco smoking and pathogenesis of
    atherosclerosis, Preventive Medicine 8:303 (1979)

25. L.L. Hawkins, Blood carbon monoxide levels as a function of daily
    cigarette consumption and physical activity, Brit.J.Ind.Med.
    33:123 (1976)

26. P.V. Cole, Comparative effects of atmospheric pollution and
    cigarette smoking on carboxyhaemoglobin levels in man, Nature
    (London) 255:699 (1975)

27. R.D. Steward, E.D. Baretta, L.R. Platte, E.B. Stewart, J.H. Kald-
    fleisch, B. van Yserloo and A.A. Rimm, Carboxyhemoglobin levels
    in American blood donors, J.Am. Med.Assn. 229:1187 (1974)

28. V. Scheidemandel and S. Daum, Carboxyhemoglobin concentrations
    of the Munich population, Munchen Med. Wochenschr. 115:109 (1973)

29. H. Schievelbein,G. Heinemann, K. Loeschenkohl, C. Troll and J.
    Schlegel, Metabolic aspects of smoking behaviour, in "Smoking
    Behaviour", R.E. Thornton,ed., Churchill Livingstone, Edinburgh-
    London-New York (1978)

30. R.J.B. Garrett and M.A. Jackson, Effect of acute smoke exposure
    on hepatic protein synthesis, J.Pharm.Exp.Ther. 209:215 (1979)

31. R.S. Gardner, D.L. Topping and P.A. Mayes, Immediate effects of
    carbon monoxide on the metabolism of chylomicron remnants by
    perfused liver, Biochim.Biophys.Res.Comm. 82:526 (1978)

32. R.E. Vestal, A.H. Norris, J.D. Tobin, H.B. Cohen, W.N. Shock and
    R. Andress, Antipyrine metabolism in man. Influence of age ,
    alcohol, caffeine and smoking, Clin.Pharm.Ther. 18:425 (1975)

33. W.S. Aronow, Effect of non-nicotine cigarettes and CO on angina,
    Circulation 61:262 (1980)

34. H. Aston, R. Stepney and J.W. Thompson, Should intake of CO be
    used as a guide to intake of other smoke constituents ? Brit.
    Med.J. 282:10 (1981)

35. U. Spohr, K. Hoffmann, W. Steck, J. Harenberg, E. Walter, N. Hen-
    gen, J. Augustin; H. Mörl, A. Koch, A. Horsch and E. Weber,
    Evaluation of smoking induced effects on sympatic,hemodynamic
    and metabolic variables with respect to plasma nicotine and
    Hb-CO levels, Atherosclerosis 33:271 (1979)

36. B.D. Rabinowitz, K. Thop, G.L. Huber and W.H. Abelmann, Acute
    hemodynamic effects of cigarette smoking in man assessed by
    systolic time intervals and echocardiography, Circulation
    60:752 (1979)

37. R.F. Isaac and M.J. Rand., Cigarette smoking and plasma levels
    of nicotine, Nature (London) 236:308 (1972)

38. A.K. Armitage, C.T. Dollery, C.F. George, T.H. Houseman, P.J.
    Lewis and D.M. Turner, Absorption and metabolism of nicotine
    by man during cigarette smoking, Brit.J.Clin.Pharm. 1:180 (1974)

39. P. Hill, Nicotine:An etiological factor for coronary heart disease,
    in "Smoking and Health.I.Modifying the Risk for the Smoker",
    E.L. Wynder, D. Hoffmann and G.B. Gori,eds., U.S.A.,DHEW (1976)

40. P.E. Cryer, M.W. Haymond, J.V. Santiago and S.D. Shah, Norepi-
    nephrine and epinephrine release and adrenergic mediation of
    smoking:Associated hemodynamic and metabolic events, New Engl.
    J.Med. 295:573 (1976)

41. R.J. Lefkewitz, Smoking, catecholamines and the heart, New Engl.
    J.Med. 295:615 (1976)

42. P. Hill and E. Wynder, Smoking and cardiovascular disease. Effect
    of nicotine on the serum epinephrine and corticoids, Am.Heart J.
    87:491 (1974)

43. A. Kershbaum, D.J. Pappajohn, S. Bellet, M. Hirabayashi and H.
    Shafiita, Effect of smoking and nicotine on adrenocortical
    secretion, J.Am.Med.Assn. 203:275 (1968)

44. O.D. Mjos and A. Ilebekk, Effects of nicotine on myocardial
    metabolism and performance in dogs, Scand.J.Clin.Lab.Invest.
    32:75 (1978)

45. L. Tachmes, R.I. Fernandez and M.A. Sackner, Hemodynamic effects
    of smoking cigarettes of high and low nicotine content, Chest
    74:243 (1978)

46. J.W. Davis and R.F. Davis, Acute effect of tobacco cigarette
    smoking on the platelet aggregate ratio, Am.J.Med.Sci. 278:139
    (1979)

47. R.H. Levine, An acute effect of cigarette smoking on platelet
    function:A possible link between smoking and arterial throm-
    bosis, Circulation 48:619 (1973)

48. R.I. Hawkins, Smoking, platelets and thrombosis, Nature(London) 236:450 (1972)

49. M.L. Bierenbaum, A.I. Fleischman, A. Stier, H. Somol and P.B. Watson, Effect of cigarette smoking upon in vivo platelet function, Thrombosis Res. 12:1051 (1978)

50. G. Grignani, G. Gamba and E. Ascari, Cigarette smoking effect on platelet function, Thrombosis and Haemostasis 37:423 (1977)

51. D.H. Lawson, J.F. Davidson and H. Jick, Oral contraceptive use and venous thromboembolism:absence of an effect of smoking , Br.Med.J. ii:729 (1977)

52. C.G. Becker and T. Dubin, Activation of factor XII by tobacco glycoprotein, J.Exp.Med. 146:457 (1977)

53. S.B. Lehrer, M.R. Wilson and J.E. Salvaggio, Immunogenic properties of tobacco smoke, J. Allergy Clin.Immunol. 62:368 (1978)

54. C.R. Minick, Immunological arterial injury in atherogenesis, Ann.N.Y.Acad.Sci. 275:210 (1976)

55. G. Fust, E. Szondy, J. Szekely, I. Nanai and S. Geno, Circulating immune complexes in vascular diseases, Atherosclerosis 29:181 (1978)

56. C.G. Becker, T. Dubin and H.P. Wiedeman, Hypersensitivity to tobacco antigen, Proc.Natl.Acad.Sci.(USA) 73:1712 (1976)

57. C.G. Becker, R. Levi and J. Zavecz, Induction of IgE antibodies to antigen isolated from tobacco leaves and from cigarette smoke condensate, Am.J.Path. 96:249 (1979)

58. P. Clopath, Immunological arterial injury in swine, in "Peptides of the Biological Fluids", H. Peters,ed.,Pergamon Press, New York (1978)

59. U. Goldbourt and J.H. Medalie, Characteristics of smokers, nonsmokers and ex-smokers among 10.000 adult males in Israel.II. Physiologic, biochemical and genetic characteristics, Am.J. Epidemiol. 105:75 (1977)

60. I. Hjermann, A. Helgeland, I. Holme, P.G. Lund-Larsen and P. Leren, The intercorrelation of serum cholesterol, cigarette smoking and body weight,The Oslo study, Acta Med.Scand. 200:479 (1976)

61. R.J. Garrison, W.B. Kannel, M. Feinleib, W.P. Castelli, P.M. McNamara and S.J. Padgett, Cigarette smoking and HDL cholesterol, Framingham offspring study, Atherosclerosis 30:17 (1978)

62. S. Heyden, G. Heiss, C. Manegold, H. Tyroler, C.G. Hames, A.G. Bartel and G. Cooper, Combined effect of smoking and coffee drinking on LDL, HDL cholesterol,Circulation 60:22 (1979)

63. K. Berg, A.-L. Børresen and G. Dahlen, Effect of smoking on serum levels of HDL Apoproteins, Atherosclerosis 34:339 (1979)

ETHANOL AND BLOOD LIPIDS IN MAN

P. Avogaro, G. Bittolo-Bon, F. Belussi and G. Cazzolato

Sub-project Atherosclerosis, National Council for

Research, Regional General Hospital, Venice, Italy

ABOUT THE RELATIONSHIP BETWEEN ALCOHOL AND BLOOD LIPIDS IN MAN

Ethanol is a powerful inducer of hyperlipemia in both animal and man. This effect is an interesting part of the confused and far from clear correlation between alcohol and cardiovascular disease. This review will refer some of the more relevant aspects of this relationship stressing both the data pointing to alcohol as a protective factor for the myocardium and the data indicating alcohol as a source of myocardial damage.

Alcohol and atherosclerosis ; experimental and clinical studies

Studies in laboratory animals have shown that an alcohol-rich diet has no effect on aortic atherosclerosis, although plasma lipid rise (1-3). Slightly increased vascular sudanophilia has been reported in rats (4). Despite its lipid raising effect, the addition of alcohol to the diet has a dose-related inhibitory effect on atherosclerosis induced by a high cholesterol diet (5).

Abbreviations used in the text: VLDL (very low density lipoproteins); LDL (low density lipoproteins); HDL (high density lipoproteins); apo (apolipoprotein); C (cholesterol); TG (triglycerides); P (protein); NC (normal controls); HED (heavy ethanol drinkers); ALC (alcoholic liver cirrhosis); NEFA (non esterified fatty acids).

Data obtained in man are difficult to interpret because of the number of different sources of alcohol and of interference from factors which cannot always be eliminated. Autoptic reviews based on large caselists (6,7) show no relationship between alcohol intake and cardiovascular pathology ("no acceleration or slowing down"). Other large autopsy caselists (8,9) report, conversely, a lower incidence of infarction and coronary disease among ethanol drinkers. Pell and D'Alonzo (10) found a much higher coronary mortality among drinkers than in a normal population ; this finding however is rendered less meaningful in the light of the fact that arterial hypertension is at least 2-3 times more frequent among drinkers.

A study of twins with different drinking habits failed to show any relation between alcohol intake and ischaemic heart disease (11). Sirtori et al. (12) reported high alcohol consumption among patients with peripheral arterial disease, but also high consumption of tobacco and carbohydrates. Recently, Castelli et al. (13), on a epidemiological basis, found an inverse correlation between alcohol intake and cardiovascular disease. Due to the high plasma HDL-C levels recorded in alcohol drinkers, it was hypothesized that this might be one factor to explain the lower incidence of ischaemic heart disease, in view of the negative correlation between this condition and HDL-C levels (14).

A recent clinical study showed an inverse correlation between the extent of coronary occlusion, angiographically demonstrated, alcohol intake and plasma HDL-C levels (15). The authors underline that alcohol should exert its "occlusion retarding effect" through mechanisms that are probably different from those of HDL-C.

Experimental studies have shown that alcohol-induced peripheral vasodilatation through direct action on the microvascular smooth muscle, reducing the response of small vessels to vasoactive substances, and interrupting or inhibiting the action of catecholamines (16). However, a recent report describes a large number of cases of atherosclerotic brain infarction during acute alcoholic intoxication in young subjects (17).

Alcohol and Plasma Lipids

Alcohol's most widespread effect on the lipid metabolism results in the accumulation of fat in the liver, but this topic will not be discussed here. This review will in fact stress the alterations of plasma lipids induced by alcohol. Although the change in plasma lipids may depend upon the effects of alcohol on the liver, the sake for clarity, we shall not go into the various effects of

different modes of intake and/or administration of alcohol, but
we shall consider the acute and chronic effects separately.

These effects have different meanings. Lipid alterations
express the impact of alcohol on the intermediate metabolism and
may therefore represent an attempt to compensate for the altered
metabolic conditions. These alterations also indicate functional
insufficiency of the liver.

Acute effects of alcohol on plasma lipids in man

In one person out of four, oral alcohol intake or intravenous administration of alcohol results in hyperlipidemia. Plasma becomes opalescent, or milky looking, within 2-3 hours of the
acute dose, and sometimes a thick layer of chylomicrons can be
skimmed off the top of the test-tube stored at 4°C. This consists
of chylomicrons which tend to rise to the top of the tube on account
of their low density. This induced hyperlipidemia last 8-12 hours,
after which plasma returns to its normal trasparency (18).

In subjects susceptible to alcohol-induced lipidemia, this
pattern recurs each time they take alcohol. The rise in blood
lipids is largely due to an increase of plasma triglycerides , and
it is interesting to note that the mean values of the increase
(128%) is close to that obtained with carbohydrate-induced lipidemia (139%) (18).

An acute dose of alcohol has little or no effect on cholesterol
(18), and its effects on NEFA are still unclear. NEFA appear to
rise during acute alcoholic intoxication (12). Nester and Hirsch
(19) found them reduced after an oral load. The effect appears to
be dose-related; NEFA fall after moderate doses of alcohol but
rise after a large amount (20), like in any state of stress.
Johanson and Laurell were the first to observe an increase in
α-lipoproteins during acute alcoholic intoxication (21). In a
previous study, involving ultracentrifugation of the individual
lipoprotein classes during an acute load, we were able to show
that the "mass" of all types (VLDL, IDL or remnants, LDL, HDL_2
and HDL_3) becomes greater after an acute alcohol dose (18,22).
This is largely because of an increase in triglycerides, and less
on account of the proteins, not only in VLDL but in all the lipoprotein classes including the high density (HDL) (18). During the
acute load, VLDL-C rises while LDL-C and HDL-C fall (18). During
the acute alcoholic intoxication an increase in apolipoprotein C
has been reported (22). In ethanol-induced hyperlipemia the with-
drawal of ethanol induces a sharp decrease of plasma triglycerides,

cholesterol, HDL-cholesterol as well as of apoB and apoA-I.
The complete normalization of the lipid plasma pattern can be
obtained in few days. When the same subjects receive 80 grams of
ethanol daily plus the usual diet a fast increase of cholesterol,
triglycerides, HDL-cholesterol, apoB and apoA-I can be observed
in a week (Fig. 1 and 2).

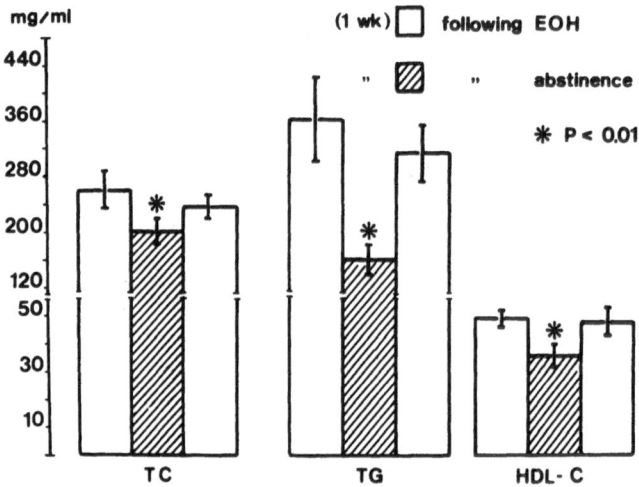

Fig. 1   Variations of cholesterol , triglycerides and HDL-choles-
         terol following one week of ethanol intake and one week
         of abstinence.

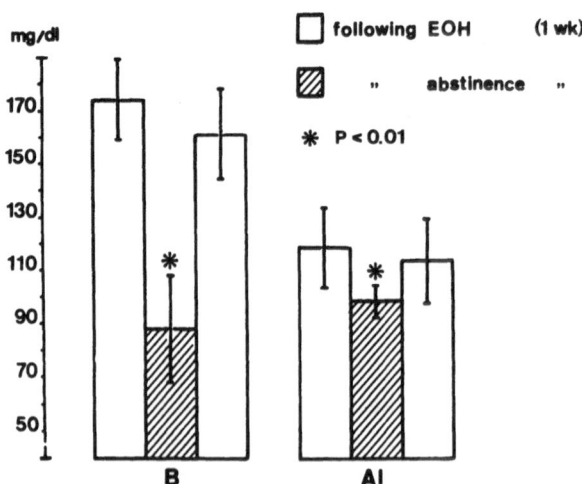

Fig. 2   Variations of apolipoproteins B and A-I following one week
         of ethanol intake and one week of abstinence.

Effects of chronic alcohol consumption on plasma lipids

    Chronic consumption of large amounts of alcohol gives rise
to a series of alterations to plasma lipids that may differ widely
depending on whether or not there are related hepatic lesions.
When chronic alcoholism is not complicated by hepatic lesions significantly higher values of triglycerides apoA-I and apoD than in
normal population are recorded while HDL-C and apoA-II are significantly reduced (23). It is worth stressing, however, that
although HDL-C and apoA-I values are not on the average higher
among habitual drinkers than in controls, alcoholics are an extremely heterogeneous population, usually with raised values for both
parameters but occasionally with very low values, like those in
alcoholic cirrhosis. In the same subjects the ratio of apoA-I to
HDL-C may be raised like the ratio of apoA-I to apoA-II (23). In
the same sub-group of alcoholics without liver impairement low
density lipoproteins, VLDL, IDL and LDL, are normal or low (23,
24); an isolated increase of HDL_2 has been reported while HDL_3 keep
in a normal range (23,24) (Table 1).

    When a chronic alcoholic patient has the complication of severe
hepatic insufficiency, there is a significant reduction of plasma
lipids (cholesterol, triglyceride , HDL-C), of the "mass" of some
lipoprotein classes (VLDL, IDL and HDL_3) and of the main apolipoproteins (A-I, A-II, B and D) (Tab.1). The A-I/A-II and HDL_2/HDL_3
ratios are significantly increased in these subjects (24).

Table 1

Plasma levels of the major lipoprotein classes in controls, heavy
ethanol drinkers and alcoholic cirrhosis.

		VLDL mg/dl	IDL mg/dl	LDL mg/dl	HDL_2 mg/dl	HDL_3 mg/dl
Controls	m	75.3	34.6	253.8	118.6	245.4
(6)	s.d.	8.3	5.4	12.8	8.9	15.6
H.E.D.	m	48.5	20.7	272.3	168.5 *	205.5
(15)	s.d.	18.3	4.5	20.3	25.7	27.2
A.L.C.	m	20.0 **	12.2 **	231.3	115.0	58.5 **
(5)	s.d.	5.2	2.7	51.7	19.5	10.2

             * = p .05 ** = p .01 vs controls

Both chronic drinkers with normal hepatic function and those with severe hepatic complications show a significant reduction in Lp(a) (23), a lipoprotein considered as an indipendent risk factor for atherosclerosis (25;26).

The low VLDL levels noted in chronic alcoholics and the resulting high HDL_2 levels are considered to be linked with the marked activation of lipoprotein lipase in adipose tissue and of post-heparin plasmatic lipolytic activity (24) observed in chronic drinkers.

The low lipid, lipoprotein and apolipoprotein values found in alcoholism complicated by cirrhosis derive from liver functional insufficiency. HDL-C, apoA-I and A-II and apoB concentrations are all significantly correlated with plasma albumin and prothrom-bin, and negatively with bilirubin; apoD is positively correlated with SGOT and SGPT levels (23)(Tab. 2).

Table 2

Correlation coefficients between some lipids and apolipoproteins and some parameters of liver function.

	Albumin	SGOT	SGPT	Prothr.	Bil.
	- correlation coefficients -				
Cholesterol	.373**	-.182	-.001	.330**	-.293*
Triglycerides	.192	-.270*	-.103	.176	-.192
HDL-C	.265*	-.122	-.021	.312**	-.279*
ApoA-I	.254*	-.040	-.153	.293*	-.315**
ApoA-II	.533**	-.210	-.079	.536**	-.599**
ApoD	.167	.423**	.338**	-.040	-.019
ApoB	.259*	.203	-.069	.226	-.197
Lp(a)	.146	-.082	.045	.311**	-.218

* = P   .01   ** = P   .001

REFERENCES

1. Nikkila, E.A., and Ollila, O., Effect of alcohol ingestion on
    experimental chicken atherosclerosis, Circ.Res. 7: 588-594
    (1959).
2. Vasdev, S.C., Charravarti, R.N., Subrahlmanyam, D., and Wahi, P.L.,
    Effect of chronic alcohol administration on blood and tissue
    lipids and hystopathology of the heart in the Rhesus monkeys,
    Indian J.Med.Res. 63: 420-425 (1973).
3. Nichols, C.W., Siperstein, M.D., Gaffey, W., Lindsay, S., and Chai-
    koff, I.L., Does the ingestion of alcohol influence the de-
    velopment of atherosclerosis in fowls?, J.Exp.Med. 103:
    465-475 (1956).
4. Gottlieb, L.S., Broitman, S.A., Vitale, J.J., and Zamcheck, N.,
    The influence of alcohol and dietary magnesium upon hypercholes-
    terolemia and atherogenesis in the rat, J.Lab.Clin.Med. 53:
    433-441 (1959).
5. Goto, Y., Hikuchi, H., Abe, K., Nagawashi, Y., Ohira, S., and Kudo,
    H., The effect of ethanol on the onset of experimental athero-
    sclerosis, Tohoku J.Exp.Med. 114: 35-43 (1974).
6. Viel, B.S., Danson, S., Slacedo, D., Dojas, P., Varela, N., and
    Alessandri, R., Alcoholism and socioeconomic status, hepatic
    damage and atherosclerosis, Arch.Intern.Med. 117: 84-91 (1966).
7. Spain, D.M., and Bradess, V.A., Sudden death from coronary athero-
    sclerosis, Arch.Intern.Med. 100: 228-231 (1957).
8. Hirst, A.E., Hadley, G.G., and Gore, I., The effect of chronic al-
    coholism and cirrhosis of the liver on atherosclerosis, Am.J.
    Med.Sci. 249: 143-149 (1965).
9. Grant, M.C., Wasserman, F., Rodensky, P.L., and Thomson, R.V.,
    The incidence of myocardial infarction in portal cirrhosis, Ann.
    Int.Med. 51: 774-779 (1959).
10. Pell, S.A., and D'Alonzo, C.A., A five-years mortality study of
    alcoholics, J.Occup.Med. 15: 120-125 (1973).
11. Myrhed, M., Alcohol consumption in relation to factors associat-
    ed with ischemic heart disease, Acta Med.Scand. 196, suppl.
    567: 8-93 (1974).
12. Sirtori, C.R., Biasi, G., Vercellio, G., Agradi, E., and Malan, E.,
    Diet lipids and lipoproteins in patients with peripheral vas-
    cular disease, Am.J.Med.Sci. 268: 325-332 (1974).
13. Castelli, W.P., Gordon, T., and Hjortland, M.C., Alcohol and blood
    lipids; the cooperative phenotyping study, Lancet ii: 53-157
    (1977).
14. Miller, C.J., and Miller, M.E., Plasma high-density-lipoprotein
    concentration and development of ischemic heart disease,
    Lancet i: 16-19 (1975).

15. Barboriak, J.J., Anderson, A.J., and Hoffmann, R.G., Interrelationship between coronary artery occlusion, high-density-lipoprotein cholesterol, and alcohol intake, J.Lab.Clin.Med. 94: 348-353 (1979).

16. Altura, B.M., Ogunkoya, A., Gebrewold, A., and Altura, B.T., Effect of ethanol on terminal arterioles and muscular venules; direct observations on the microcirculation, J. Cardiovascular Pharmocol. 1: 97-113 (1979).

17. Hillbom, M., and Kaste, M., Does ethanol intoxication promote brain infarction in young adults?, Lancet ii: 1181-1183 (1978).

18. Avogaro, P., and Cazzolato, G., Changes in the composition and Physico-chemical characteristics of serum lipoproteins during ethanol-induced lipemia in alcoholic subjects, Metabolism 24: 1231-1242 (1975).

19. Nestel, P.J., and Hirsch, E.Z., Clinical and experimental mechanism of alcohol-induced hypertriglyceridemia, J.Lab.Clin.Med. 66: 357-365 (1965).

20. Baraona, E., and Lieber, C.S., Effects of ethanol on lipid metabolism, J.Lipid.Res. 20: 289-315 (1979).

21. Johanson, B.G., and Laurell, C.B., Disorders of serum α-lipoproteins after alcoholic intoxication, Scand.J.Clin.Lab.Invest. 23: 231-233 (1979).

22. Avogaro, P., Cazzolato, G., Bittolo Bon, G., and Belussi, F., Ethanol lipemia. A doubtful risk factor. In: Atherosclerosis IV, G. Schettler, Y. Goto, Y. Hata, and G. Klose, eds., Springer-Verlag, Berlin, pp. 658-664 (1977).

23. Avogaro, P., Bittolo Bon, G., Cazzolato, G., and Belussi, F., Lipoproteins and apolipoproteins in chronic alcoholics and in alcoholic liver cirrhosis, Recent Adv. in Clinical Nutr. i, John Libbey and Co., London (in press).

24. Nilsson-Ehle, P., Alcohol induced alterations in lipoproteins lipase activity and plasma lipoproteins. In: Metabolic effects of alcohol, P. Avogaro, C.R. Sirtori, and E. Tremoli, eds., Elsevier/North Holland, pp. 175-186 (1979).

25. Walton, K.W., Hitchens, J., Magnani, H.N., and Khan M., A study of methods of identification and estimation of Lp(a) lipoprotein and of its significance in health, hyperlipidemia and atherosclerosis, Atherosclerosis 20: 223-234 (1974).

26. Kostner, G.M., Avogaro, P., Cazzolato, G., Marth, E., Bittolo Bon, G., and Quinci, G.B., Lipoprotein Lp(a) and the risk for myocardial infarction, Atherosclerosis (in press).